Introduction to the Analysis of Electromechanical Systems

IEEE Press
445 Hoes Lane
Piscataway, NJ 08854

Introduction to the Analysis of Electromechanical Systems

Paul C. Krause
Oleg Wasynczuk
Timothy O'Connell

IEEE Press Series on Power and Energy Systems
Ganesh Kumar Venayagamoorthy, Series Editor

IEEE PRESS
WILEY

Published by John Wiley & Sons, Inc., Hoboken, New Jersey.
Published simultaneously in Canada.

Library of Congress Cataloging-in-Publication Data is applied for

Hardback: 9781119829997

Cover Design: Wiley

MIX
Paper from responsible sources
FSC www.fsc.org
FSC® C013604

Contents

Preface

In the past, a course in electric machines has generally been the first course in the study of electromechanical systems; however, this area now includes electric machines, power electronics, and electric drives, in addition to electric power systems. This has caused some concern as to the most appropriate first course. Although the analysis of electric machinery is fundamental to electromechanical systems, it may not be the most effective way to introduce this area to the student who is trying to decide on a career path. Some schools have instituted a survey course; however, this often lacks the depth to provide the student an introduction to the analysis common to electromechanical systems or to provide an analytical foundation on which to build in follow-on courses. An introductory course that establishes a useful analytical foundation seems appropriate. This book is an attempt to fulfill this need.

Tesla's rotating magnetic field is the basis of reference frame theory used in the analysis of electromechanical systems, since all known mathematical transformations that are used in this area are contained in the expression of Tesla's rotating magnet field knowledge of this is very important to the engineer working in machines, drives, power and power systems stability. This approach provides an analytical means of positioning the stator and rotor poles on the phasor diagram, thereby providing a straightforward and instructive illustration of the operation of the electromechanical system. Tesla not only invented the ac machine; his work is instrumental in analyzing and visualizing its operation.

Although reference frame theory in Chapter 2 should be covered, the reference frame derivations in later chapters may be deemphasized. Most of the material in Chapter 1 is common to all disciplines. The subsequent chapters contain material fundamental to the areas of electric machines, power electronics, electric drives, and electric power systems. Material that is somewhat more advanced is included in most cases. This allows the instructor to choose between a brief or a more in-depth coverage.

Electric machines are covered in Chapters 2 and 3, focusing on the analysis of the dc, round-rotor permanent-magnet ac, and induction machines. Power electronics is covered in Chapter 4. The dc, the brushless dc, and the induction motor drives are each covered in Chapter 5. The synchronous generator and several other aspects of electric power system engineering are covered in Chapter 6. Hopefully, this book will help the student in deciding if he or she would want to pursue advanced study in electromechanical systems.

This book has several uses, it can be used as a one- to three-hour course in schools that offer only one course in the power and drives area. The material can be taught at the sophomore level or first-semester junior level. For schools that have a power program, the course could be used as the introductory course.

Obviously, there are other aspects that would be appropriate for an introductory text; however, the choices we have made seem to be representative at this time. Nevertheless, as the area continues to evolve through the twenty-first century, so must an introductory text.

We would like to acknowledge Dr Brett Robbins of PCKA for developing the drawings and formatting the computer traces and Chris Ramsey who typed the draft of the text.

West Lafayette, Indiana, USA
April 2021

Paul C. Krause, Oleg Wasynczuk, and
Timothy O'Connell

About the Authors

PAUL C. KRAUSE, PhD, retired after 39 years as a professor at Purdue University School of Electrical and Computer Engineering. He founded P.C. Krause & Associates in 1983. He is a Life Fellow of IEEE and has authored or co-authored over 100 technical papers and five textbooks on electric machines. He was the 2010 recipient of the IEEE Nikola Tesla Award.

OLEG WASYNCZUK, PhD, is a Professor of Electrical and Computer Engineering at Purdue University. He has authored or co-authored over 100 technical papers and four textbooks on electric machines. He is a Fellow of IEEE and was the 2008 recipient of the IEEE Cyril Veinott Award. He also serves as Chief Technical Officer of P.C. Krause & Associates.

TIMOTHY O'CONNELL, PhD, is a Senior Lead Engineer at P.C. Krause & Associates and an Adjunct Professor of Electrical and Computer Engineering at the University of Illinois at Urbana-Champaign. He is a Senior Member of IEEE, and an Associate Editor of the IEEE *Transactions on Aerospace and Electronic Systems*. He has co-authored over 20 technical papers and three textbooks on electric machines and has co-edited a book on electrified aircraft propulsion.

1

Basic System Analysis

1.1 Introduction

The twentieth century began with the electric power industry in its infancy; Thomas Edison and Nikola Tesla were locked in battle with Edison advocating direct current (dc) and Tesla alternating current (ac). The century ended with the electric power industry expanding rapidly from the traditional power generation, transmission, and utilization into propulsion of air, ground, and sea transportation. The advent of the computer and the silicon-controlled rectifier in the mid-1900s brought about an expansion of the power area to include the smart-grid, microgrids, efficient and robust electric drives, more electric aircraft, ships, and land vehicles. A growth which is likely to continue into the foreseeable future.

Before the advent of the computer, engineers were essentially limited to steady-state analysis and therefore unable to conveniently deal with the analytical challenges of the expanding power industry. This chapter sets forth some of the basic concepts and analysis tools that are part of the present-day power and electric drives area. Although not inclusive, the material covered in this chapter is representative and common to most disciplines of the power area.

1.2 Phasor Analysis and Power Calculations

Since the early twentieth century, we have lived in an ac world. Thanks to George Westinghouse and Nikola Tesla, power systems are predominately ac; power is generated by large ac generators, transmitted by high-voltage transmission lines, and transformed to a low voltage and distributed to homes and factories. The evolution of the ac power system brought about many engineering challenges and, as we look back, it is difficult to comprehend how these problems were solved without a computer. Even steady-state ac-circuit analysis posed a problem until

Introduction to the Analysis of Electromechanical Systems, First Edition.
Paul C. Krause, Oleg Wasynczuk, and Timothy O'Connell.

the early 1900s when Charles Stienmetz, who was a less flamboyant colleague of Edison and Tesla, came up with the concept of what is now known as phasors. Some may consider the phasor a casualty of the computer age along with the slide rule. It is, however, still a very useful means for understanding and portraying the steady-state performance of electric machines, power systems, and electric drives. Moreover, the phasor concept provides a means of visualizing sinusoidal variations from different frames of reference, and in Chapter 2 we will find that the voltage and current phasors combined with Tesla's rotating magnetic field provides a straightforward means of analyzing and portraying the steady-state operation of ac machines.

The phasor can be established by expressing a steady-state sinusoidal variable as

$$F_a(t) = F_p \cos \theta_{ef} \tag{1.1}$$

where the a subscript is used here to denote sinusoidal quantities. The sinusoidal variations are expressed as cosines, capital letters are used to denote steady-state quantities, and F_p is the peak value of the sinusoidal variation. Generally, F or f represents voltage (V or v) or current (I or i) in circuit analysis, but it could be any sinusoidal variable. For steady-state conditions, θ_{ef} may be written as

$$\theta_{ef}(t) = \omega_e t + \theta_{ef}(0) \tag{1.2}$$

where ω_e is the electrical angular velocity in rad/s (2π times the frequency) and $\theta_{ef}(0)$ is the time-zero position of the electrical variable. Substituting (1.2) into (1.1) yields

$$F_a(t) = F_p \cos\left[\omega_e t + \theta_{ef}(0)\right] \tag{1.3}$$

Now, Euler's identity is

$$e^{j\alpha} = \cos\alpha + j\sin\alpha \tag{1.4}$$

and since we are expressing the sinusoidal variation as a cosine, (1.3) may be written as

$$F_a(t) = \mathrm{Re}\left\{F_p e^{j[\omega_e t + \theta_{ef}(0)]}\right\} \tag{1.5}$$

where Re is shorthand for the "real part of." Equations (1.3) and (1.5) are equivalent. Let us rewrite (1.5) as

$$F_a(t) = \mathrm{Re}\left\{F_p e^{j\theta_{ef}(0)} e^{j\omega_e t}\right\} \tag{1.6}$$

We need to take a moment to define what is referred to as the root-mean-square (rms) of a sinusoidal variation. In particular, the rms value is defined as

$$F = \left(\frac{1}{T}\int_0^T F_a^2(t)dt\right)^{\frac{1}{2}} \tag{1.7}$$

where F is the rms value of $F_a(t)$ and T is the period of the sinusoidal variation. It is left to the reader to show that the rms value of (1.3) is $F_p/\sqrt{2}$. Therefore, we can express (1.6) as

$$F_a(t) = \text{Re}\left[\sqrt{2}Fe^{j\theta_{ef}(0)}e^{j\omega_e t}\right] \tag{1.8}$$

By definition, the phasor representing $F_a(t)$, which is denoted with a raised tilde, is

$$\tilde{F}_a = Fe^{j\theta_{ef}(0)} \tag{1.9}$$

which is a complex number. The reason for using the rms value as the magnitude of the phasor will be addressed later in this section. Equation (1.6) may now be written as

$$F_a(t) = \text{Re}\left[\sqrt{2}\tilde{F}_a e^{j\omega_e t}\right] \tag{1.10}$$

A shorthand notation for (1.9) is

$$\tilde{F}_a = F\underline{/\theta_{ef}(0)} \tag{1.11}$$

Equation (1.11) is commonly referred to as the *polar form* of the phasor. The *Cartesian* form is

$$\tilde{F}_a = F\cos\theta_{ef}(0) + jF\sin\theta_{ef}(0) \tag{1.12}$$

When using phasors to calculate steady-state voltages and currents, we think of the phasors as being stationary at $t = 0$; however, we know from (1.10) that a phasor is related to the instantaneous value of the sinusoidal quantity it represents. In other words, the real projection of the phasor $\tilde{F}_a$ rotating counterclockwise at ω_e is the instantaneous value of $F_a(t)/\sqrt{2}$. Thus, with $\theta_{ef}(0) = 0$ in (1.3)

$$F_a(t) = \sqrt{2}F\cos\omega_e t \tag{1.13}$$

the phasor representing (1.13) is

$$\tilde{F}_a = Fe^{j0} = F\underline{/0^\circ} = F + j0 \tag{1.14}$$

For

$$\begin{aligned} F_a(t) &= \sqrt{2}F\sin\omega_e t \\ &= \sqrt{2}F\cos(\omega_e t - 90^\circ) \end{aligned} \tag{1.15}$$

the phasor is

$$\tilde{F}_a = Fe^{-j\pi/2} = F\underline{/-90^\circ} = 0 - jF \tag{1.16}$$

We will use degrees and radians interchangeably when expressing phasors. Although there are several ways to arrive at (1.16) from (1.15), it is helpful to ask yourself where must the rotating phasor be positioned at time zero so that, when it rotates counterclockwise at ω_e, its real projection is $(1/\sqrt{2})F_p \sin\omega_e t$? It follows that a phasor of amplitude F positioned at 90° represents $-\sqrt{2}F\sin\omega_e t$.

In other words, we are viewing a sinusoidal variation as the real projection in the real-imaginary plane of a rotating line equal in magnitude to the positive peak value ($\sqrt{2}F$) of the variation and rotating at the electrical angular velocity of the sinusoidal variation. Since we are dealing with a steady-state variation, we can stop the rotation at any time and view it as a fixed line, but knowing full well that it, in fact, represents a sinusoidal variations and to represent the sinusoidal variation we must rotate it counterclockwise at ω_e and take the real projection. Please understand that if we ran at ω_e in unison with the rotating $\sqrt{2}F$ line it would appear as a constant to us. Therefore, this is no different than stopping the phasor at some arbitrary time zero; but realizing that it actually represents a sinusoidal variation. We'll talk more about this important aspect as we go along. See Example 1A.

To show the facility of the phasor in the analysis of steady-state performance of ac circuits and devices, it is useful to consider a series circuit consisting of a resistance, an inductance L and a capacitance C. Thus, using uppercase letters to indicate steady-state variables

$$V_a = RI_a + L\frac{dI_a}{dt} + \frac{1}{C}\int I_a dt \tag{1.17}$$

Throughout the text, we will use either R or r to represent resistance. For steady-state operation, let

$$V_a = \sqrt{2}V\cos[\omega_e t + \theta_{ev}(0)] \tag{1.18}$$

$$I_a = \sqrt{2}I\cos[\omega_e t + \theta_{ei}(0)] \tag{1.19}$$

where we have dropped the functional notation, and the subscript a helps to distinguish the instantaneous value from the rms value of the steady-state variables. The steady-state voltage equation may be obtained by substituting (1.18) and (1.19) into (1.17), whereupon we can write

$$\begin{aligned}\sqrt{2}V\cos[\omega_e t + \theta_{ev}(0)] = {} & R\sqrt{2}I\cos[\omega_e t + \theta_{ei}(0)] \\ & + \omega_e L\sqrt{2}I\cos\left[\omega_e t + \frac{1}{2}\pi + \theta_{ei}(0)\right] \\ & + \frac{1}{\omega_e C}\sqrt{2}I\cos\left[\omega_e t - \frac{1}{2}\pi + \theta_{ei}(0)\right]\end{aligned} \tag{1.20}$$

The second term on the right-hand side of (1.20), which is $L\frac{dI_a}{dt}$, can be written

$$\omega_e L\sqrt{2}I\cos\left[\omega_e t + \frac{1}{2}\pi + \theta_{ei}(0)\right] = \omega_e L\text{Re}\left[\sqrt{2}Ie^{j\frac{1}{2}\pi}e^{j\theta_{ei}(0)}e^{j\omega_e t}\right] \tag{1.21}$$

Since $\widetilde{I}_a = Ie^{j\theta_{ei}(0)}$, from (1.21), we can write

$$L\widetilde{\frac{dI_a}{dt}} = \omega_e L e^{j\frac{1}{2}\pi}\widetilde{I}_a \tag{1.22}$$

Since $e^{j\frac{1}{2}\pi} = j$, (1.22) may be written

$$L\widetilde{\frac{dI_a}{dt}} = j\omega_e L\widetilde{I}_a \tag{1.23}$$

If we follow a similar procedure, we can show that

$$\frac{1}{C}\widetilde{\int I_a dt} = -j\frac{1}{\omega_e C}\widetilde{I}_a \tag{1.24}$$

Differentiation of a steady-state sinusoidal variable rotates the phasor counterclockwise by $\frac{1}{2}\pi$ or j; integration rotates the phasor clockwise by $\frac{1}{2}\pi$ or $-j$.

The steady-state voltage equation given by (1.20) can now be written in phasor form as

$$\widetilde{V}_a = \left[R + j\left(\omega_e L - \frac{1}{\omega_e C}\right)\right]\widetilde{I}_a \tag{1.25}$$

We can express (1.25) compactly as

$$\widetilde{V}_a = Z\widetilde{I}_a \tag{1.26}$$

where the impedance, Z, is a complex number; it is not a phasor. It may be expressed as

$$Z = R + j(X_L - X_C) \tag{1.27}$$

where $X_L = \omega_e L$ is the inductive reactance and $X_C = \frac{1}{\omega_e C}$ is the capacitive reactance. We should be careful here. Some prefer to write (1.27) as $R + jX$ where X is $X_L + X_C$ and let X_C be negative. This is essentially a matter of choice and does not change the end result. We will deal primarily with X_L and not X_C, therefore, this will have little impact on our work; nevertheless, since some authors will use a negative X_C we should make the reader aware of this difference.

It is appropriate to discuss the notation that will be used throughout the text. When an equation is written with the variables in lowercase letters it is valid for transient and steady state. If the variables are written with uppercase letters as in (1.17), the equation is a function of time and valid for instantaneous steady-state conditions. Equations (1.26) and (1.27) are phasor equations representing steady-state sinusoidal variables and are written in uppercase letters with an over tilde.

1.2.1 Power and Reactive Power

The instantaneous steady-state power is

$$\begin{aligned} P &= V_a I_a \\ &= \sqrt{2}V\cos[\omega_e t + \theta_{ev}(0)]\sqrt{2}I\cos[\omega_e t + \theta_{ei}(0)] \end{aligned} \tag{1.28}$$

where V and I are rms values. After some manipulation, we can write (1.28) as

$$P = VI\cos[\theta_{ev}(0) - \theta_{ei}(0)] + VI\cos[2\omega_e t + \theta_{ev}(0) + \theta_{ei}(0)] \tag{1.29}$$

The instantaneous steady-state power given by (1.29) varies about an average value at a frequency of $2\omega_e$. That is, the second term of (1.29) has a zero average value and the average power P_{ave} may be written

$$P_{\text{ave}} = \left|\tilde{V}_a\right| \left|\tilde{I}_a\right| \cos[\theta_{ev}(0) - \theta_{ei}(0)] \tag{1.30}$$

where $\left|\tilde{V}_a\right|$ and $\left|\tilde{I}_a\right|$ are V and I, respectively, which are the magnitudes of the phasors (rms value), $\theta_{ev}(0) - \theta_{ei}(0)$ is referred to as the *power factor angle* φ_{pf}, and $\cos[\theta_{ev}(0) - \theta_{ei}(0)]$ is the *power factor*. Power is in watts. If current is assumed positive in the direction of voltage drop, then (1.30) is positive if power is consumed and negative if power is generated. It is interesting to point out that in going from (1.28) to (1.29), the coefficient of the two right-hand terms is $\frac{1}{2}(\sqrt{2}V\sqrt{2}I)$ or one-half the product of the peak values of the sinusoidal variables. Therefore, it was considered more convenient to use the rms values for the phasors, whereupon average steady-state power could be calculated by the product of the magnitude of the voltage and current phasors as given by (1.30).

The reactive power is defined as

$$Q = \left|\tilde{V}_a\right| \left|\tilde{I}_a\right| \sin\left[\theta_{ev}(0) - \theta_{ei}(0)\right] \tag{1.31}$$

The units of Q are var (volt–ampere reactive). An inductance is said to absorb reactive power where the current lags the voltage by 90° and Q is positive. In the case of a capacitor, where the current leads the voltage by 90°, Q is supplied and is negative. Actually, Q is a measure of the interchange of energy stored in the electric (capacitor) and magnetic (inductor) fields. However, unlike instantaneous real power, the average value of instantaneous reactive power is zero. We'll talk more about reactive power later.

Example 1A *Phasor analysis*

The parameters of a series RLC circuit are $R = 6\ \Omega$, $L = 20$ mH, $C = 1\times10^3\ \mu$F. The 60-Hz applied voltage is $V_a = 155.6\cos\omega_e t$. Calculate $\tilde{I}_a$, P_{ave}, Q and draw the phasor diagram and the sinusoidal variations as viewed running at counterclockwise with the phasor representing maximum V_a ($\sqrt{2}V\underline{/0°}$). From the expression of V_a

$$\tilde{V}_a = 110\underline{/0°}\ \text{V} \tag{1A.1}$$

Now, $\omega_e = 2\pi f = 2\pi \times 60 = 377$ rad/s and

$$Z = R + j(X_L - X_C)$$
$$= R + j\left(\omega_e L - \frac{1}{\omega_e C}\right)$$
$$= 6 + j\left(377 \times 20 \times 10^{-3} - \frac{1}{377 \times 1 \times 10^{-3}}\right) = 7.73\angle 39.1^\circ\ \Omega \qquad (1A.2)$$

$$\tilde{I}_a = \frac{\tilde{V}_a}{Z} = \frac{110\angle 0^\circ}{7.73\angle 39.1^\circ} = 14.2\angle -39.1^\circ \text{ A} \qquad (1A.3)$$

$$P_{ave} = \left|\tilde{V}_a\right| \left|\tilde{I}_a\right| \cos\varphi_{pf} \qquad (1A.4)$$

where

$$\varphi_{pf} = \theta_{ev}(0) - \theta_{ei}(0)$$
$$= 0 - (-39.1^\circ) = 39.1^\circ \qquad (1A.5)$$

$$P_{ave} = 110 \times 14.2 \cos 39.1^\circ$$
$$= 1212.2 \text{ W} \qquad (1A.6)$$

$$Q = \left|\tilde{V}_a\right| \left|\tilde{I}_a\right| \sin\varphi_{pf}$$
$$= 110 \times 14.2 \sin 39.1^\circ = 985.1 \text{ vars} \qquad (1A.7)$$

The phasor diagram is shown in Figure 1A.1.

SP1.2.1 Express the instantaneous steady-state power for Example 1A. [Substitute into (1.29)]

SP1.2.2 Redraw the phasor diagram shown in Figure 1A.1 showing $jX_L\tilde{I}_a$ and $-jX_C\tilde{I}_a$ as individual voltages. [Show $jX_L\tilde{I}_a$ and then from the terminus of $jX_L\tilde{I}_a$ show $-jX_C\tilde{I}_a$]

Figure 1A.1 Phasor diagram.

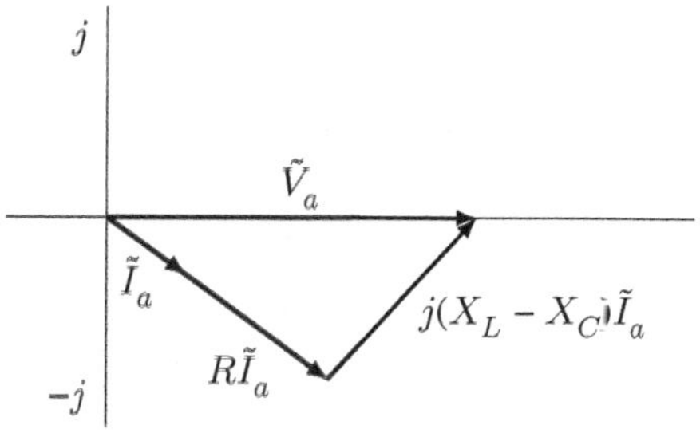

SP1.2.3 We know that $P_{\text{ave}} = |\widetilde{I}|^2 R$ does $Q = |\widetilde{I}_a|^2 X_L - |\widetilde{I}_a|^2 X_C$? [Yes]

SP1.2.4 If $\widetilde{V} = 1\underline{/0°}$ V and $\widetilde{I} = 1\underline{/180°}$ A in the direction of the voltage drop, calculate Z and P_{ave}. Is power generated or consumed? [$(-1 + j0)$ Ω, 1 W, generated]

SP1.2.5 Express the instantaneous power for 60 Hz voltage, $\widetilde{V}_a = 1\underline{/0°}$, applied to a resistive circuit, $\widetilde{I}_a = 1\underline{/0°}$. [$1 + \cos 754t$]

SP1.2.6 Repeat SP1.2.5 for (a) an inductance, $\widetilde{I}_a = I_L\underline{/-90°}$ and (b) a capacitance, $\widetilde{I}_a = I_C\underline{/90°}$. [(a) $I_L \cos(754t - 90°)$, (b) $I_C \cos(754t + 90°)$]

1.3 Elementary Magnetic Circuits

Electric machines and transformers, which are the backbone of the power industry, are electromagnetic systems. Therefore, magnetic circuits and magnetic coupling play a major role in power and drives systems, and it is necessary to establish the principles of magnetic systems sufficiently to convey the basic operation of the electromagnetic devices considered in later chapters. We will attempt to do this without becoming too involved.

An elementary magnetic circuit is shown in Figure 1.1. It consists of a ferromagnetic member (core) with a coil of wire of N turns wound on it and an air gap of length x. The ferromagnetic member could be iron, nickel, cobalt, or steel, for example. The voltage equation of the electric circuit may be written

$$v = ri + e \tag{1.32}$$

where r is the total resistance of the circuit, v is the source voltage, e is the voltage induced in the coil according to Faraday's law, and i is the current flowing in the circuit. The current flowing through the coil causes a magnetomotive force (mmf), which produces flux in the magnetic circuit denoted as Φ_m and Φ_l in Figure 1.1,

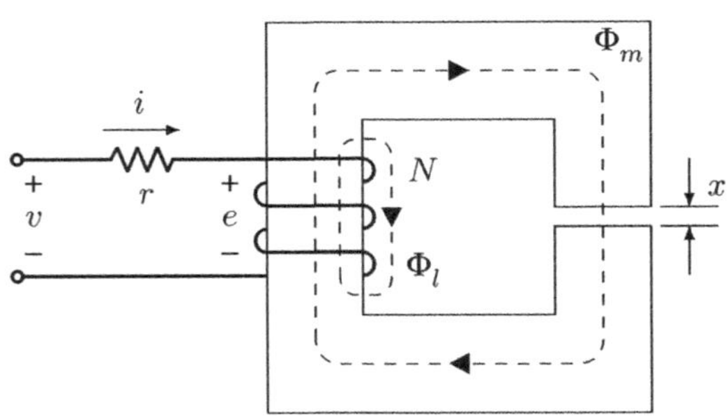

Figure 1.1 Elementary magnetic circuit with one air gap.

much as an electromotive force (emf) or source voltage produces current in an electric circuit.

There are arrows associated with the dashed lines representing the flux paths in Figure 1.1. These arrows indicate the assumed positive direction of flux which is determined from the assumed positive direction of current by the so-called "right-hand" rule. If you grasp the coil with your right hand with your fingers in the assumed direction of positive current flow around the coil, your thumb will point in the direction of positive flux. Or, imagine grasping a turn of the coil with thumb in the assumed direction of positive current. If you ungrasp your fingers, they will point in the direction of positive flux.

The total flux, Φ, that travels through (links) all of the turns N is

$$\Phi = \Phi_l + \Phi_m \tag{1.33}$$

where Φ_l is the equivalent flux that links all the turns of the coil but does not traverse the ferromagnetic member and Φ_m is referred to as the magnetizing flux that transverses the ferromagnetic member and links all of the turns of the coil. The leakage flux Φ_l is shown by one streamline in Figure 1.1, it represents the aggregate of the flux that occurs around each wire of the coil, traveling partially in the ferromagnetic material and partially in air.

The concept of magnetic poles can be used to advantage to explain the operation of electromechanical devices. The poles can be established by considering the magnetic circuit shown in Figure 1.1. In particular, to locate the north and south poles of an electromechanical device for the assumed positive direction of coil current, place yourself on the member that has the coil (windings). Use the right-hand rule to establish the direction of positive flux Φ_m for the assumed positive direction of coil current, where positive flux issues from the member with the coil into the air is the assumed north pole. The south pole is where the positive flux returns from the air to the member with the coil. In the case of the magnetic circuit shown in Figure 1.1, with the assumed direction of positive current, a north pole exists over the upper face of the air gap and a south pole over the lower face.

A magnetically linear circuit behaves much as a resistive electric circuit. According to Ohm's law, the current i in a resistive circuit is equal to the applied emf or applied voltage divided by the resistance r. In the case of a magnetic circuit, the total flux Φ is equal to the mmf, which is in ampere turns Ni, divided by the equivalent reluctance $\mathcal{R}$ of the magnetic circuit. Thus, in (1.33)

$$\Phi_l = \frac{Ni}{\mathcal{R}_l} \tag{1.34}$$

$$\Phi_m = \frac{Ni}{\mathcal{R}_i + \mathcal{R}_g} \tag{1.35}$$

where $\mathcal{R}_l$ is the reluctance of the leakage flux path, $\mathcal{R}_i$ is the reluctance of the ferromagnetic member, and $\mathcal{R}_g$ is the reluctance of the air gap. Although the leakage reluctance is generally determined by test or an involved calculation, $\mathcal{R}_i$ and $\mathcal{R}_g$ may be calculated from

$$\mathcal{R} = \frac{\ell}{\mu_r \mu_0 A} \tag{1.36}$$

where ℓ is the length of the flux path, A is the cross sectional area of the flux path, μ_0 is the permeability of free space ($4\pi \times 10^{-7}$ Wb/A m or H/m), and μ_r is the permeability relative to free space. The units of reluctance are (henry^{-1}) or H^{-1}. In the case of the air gap, the relative permeability is considered to be unity ($\mu_{rg} \approx 1$), for the ferromagnetic member typical μ_{ri} values may be as high as 4000 depending upon the type of ferromagnetic material. It follows that the reluctance of the ferromagnetic member is much less than the reluctance of either the reluctance of the air gap or the reluctance of the leakage path since it is partially in air. In fact, the reluctance of the ferromagnetic member is generally neglected when an air gap is present as in the case of an electric machine. The magnetic equivalent circuit shown in Figure 1.2 may be helpful to visualize the flux paths of the magnetic system shown in Figure 1.1.

Before proceeding, there are a couple things that we should talk about. We have arrived at (1.34) and (1.35) by exploiting the similarities between a resistive electric circuit and a linear magnetic circuit. Although this approach is straightforward and easy to follow, let's take a minute to apply basic laws of magnetically linear systems that we learned in early physics courses to justify (1.34) and (1.35). Ampere's law states that the line integral of the field intensity (field strength) H about a closed path is equal to the net current enclosed within the closed path of integration. Now, for a two dimensional magnetically linear system

$$B = \mu H \tag{1.37}$$

where B is the flux density, μ is the permeability, and H is the field strength. If we assume the flux Φ is uniform over the cross-sectional surface area of the magnetic path, then

$$B = \frac{\Phi}{A} \tag{1.38}$$

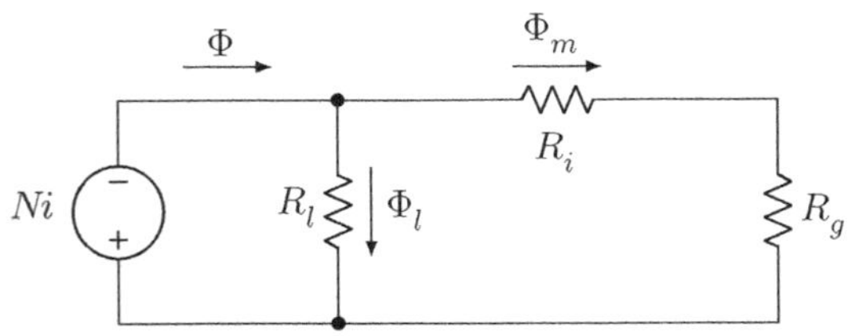

Figure 1.2 Magnetic equivalent circuit for magnetic system shown in Figure 1.1.

where A is the cross-sectional area. Thus, H may be expressed

$$H = \frac{\Phi}{A\mu} \tag{1.39}$$

Now, if H is integrated over the closed path through iron and air gap that encloses the total turns of the coil and assuming H is the same over this path, then the mmf, or Ni is

$$Ni = \int_0^{\ell} \frac{\Phi}{A\mu} d\xi \tag{1.40}$$

and

$$Ni = \frac{\Phi}{A\mu}\ell \tag{1.41}$$

Thus Φ may be expressed

$$\begin{aligned}\Phi &= \frac{Ni}{\frac{\ell}{A\mu_r\mu_0}} \\ &= \frac{Ni}{\mathcal{R}}\end{aligned} \tag{1.42}$$

where we have substituted $\mu_r\mu_0$ for μ. We now understand the difficulty in determining the reluctance to the leakage flux; however, the reluctance of the iron and air gap can be calculated quite accurately for a magnetically linear system.

We now see that (1.34) is obtained by applying Ampere's law about the leakage flux Φ_ℓ path and (1.35) is obtained by applying Ampere's law about the magnetizing flux Φ_m path. In regard to (1.35), since for a magnetically linear system the reluctance of an air gap is much larger than the reluctance of the iron, the mmf drop across the iron is generally neglected and (1.35) could be written

$$\Phi_m = \frac{Ni}{\mathcal{R}_g} \tag{1.43}$$

where the total mmf is dropped across the air gap which is commonly referred to as the air-gap mmf. We, however, will not neglect $\mathcal{R}_i$ in the following development for the sake of completeness.

The total flux linking all the turns of the coil may be expressed from (1.33) as

$$\begin{aligned}\lambda &= N\Phi \\ &= N(\Phi_l + \Phi_m)\end{aligned} \tag{1.44}$$

Substituting (1.34) and (1.35) into (1.44) yields the flux linkage as

$$\lambda = \left(\frac{N^2}{\mathcal{R}_l} + \frac{N^2}{\mathcal{R}_i + \mathcal{R}_g}\right) i \tag{1.45}$$

It is now time to get back to the electric circuit. Faraday's law tells us that the voltage induced in the coil, e in Figure 1.1, may be expressed

$$e = \frac{d\lambda}{dt} \tag{1.46}$$

Note that the coil resistance has been lumped outside the coil in the circuit shown in Figure 1.1. If we now substitute (1.45) into (1.46) and then substitute the result into (1.32) for e, we obtain

$$v = ri + \frac{d}{dt}\left[\left(\frac{N^2}{\mathcal{R}_l} + \frac{N^2}{\mathcal{R}_i + \mathcal{R}_g}\right) i\right] \tag{1.47}$$

Since we are dealing exclusively with a magnetically linear system, the reluctances are assumed to be constant; therefore, (1.47) becomes

$$v = ri + L\frac{di}{dt} \tag{1.48}$$

where L is the self-inductance which is

$$\begin{aligned} L &= \frac{N^2}{\mathcal{R}_l} + \frac{N^2}{\mathcal{R}_i + \mathcal{R}_g} \\ &= L_l + L_m \end{aligned} \tag{1.49}$$

where L_l is the leakage inductance and L_m is the magnetizing inductance. Since inductance (units of H) is turns (no units) squared divided by reluctance, the unit of reluctance is H^{-1}. Moreover, the use of an inductance in the analysis signifies that a magnetically linear system is being assumed.

Let us rewrite (1.45) as

$$\lambda = Li \tag{1.50}$$

Here we see a linear relationship between flux linkages and current. The λ versus i relationship shown in Figure 1.3 is often referred to as the magnetization curve.

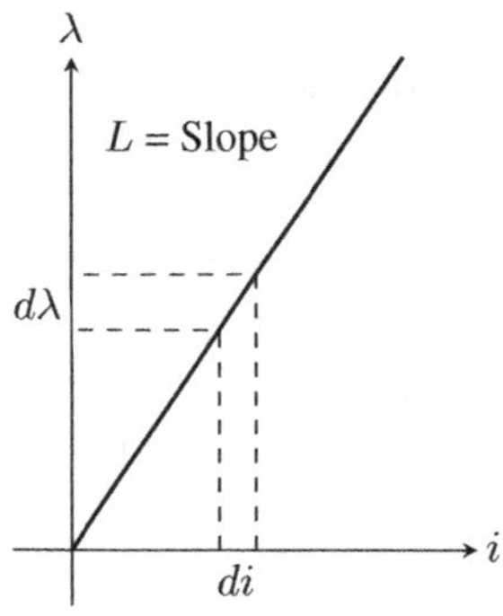

Figure 1.3 λi characteristic of a magnetically linear system.

1.3.1 Field Energy and Coenergy

Since power is the rate of energy transfer, the energy entering the magnetic field from the electrical system (Figure 1.1) may be expressed

$$W_e = \int ei\,dt \tag{1.51}$$

As previously mentioned, the resistance of the coil is lumped external to the coil; therefore, W_e is the energy transferred only to the field of the magnetic system from the electrical system. Thus, the energy stored in the field which is referred to as the field energy is

$$\begin{aligned} W_f &= W_e \\ &= \int i\frac{d\lambda}{dt}dt = \int i\, d\lambda \\ &= \frac{1}{2}Li^2 \end{aligned} \tag{1.52}$$

Therefore, the energy stored in the field (W_f) for a given λ or i is the area to the left of the magnetization curve shown in Figure 1.2. Also, since, for the magnetic system shown in Figure 1.1, $L_l \ll L_m$ and since $\mathcal{R}_i \ll \mathcal{R}_g$, most of the field energy is stored in the air gap.

The area to the right of the magnetization curve is referred to as the coenergy; it is expressed as

$$W_c = \int \lambda\, di \tag{1.53}$$

It has no physical meaning; however, it is often used as a convenience in expressing the force and torque, since it is easier to work with λ in terms of i rather than i in terms of λ. Also, we see from Figure 1.3 that

$$\lambda i = W_f + W_c \tag{1.54}$$

Clearly, $W_f = W_c$ for a magnetically linear system.

We will not take saturation into account in our analysis in this text. To do so would greatly complicate our work compared to treating all electromechanical devices as magnetically linear and little is sacrificed in portraying their salient operating characteristics. Nevertheless, magnetically nonlinear systems should be mentioned in passing. In this regard, Figure 1.4 is a plot of the flux linkage λ versus the current i for a magnetically nonlinear system. The actual characteristic is dependent on the type of ferromagnetic material. The knee of the curve occurs due

Figure 1.4 λ versus i for a magnetically nonlinear magnetic system.

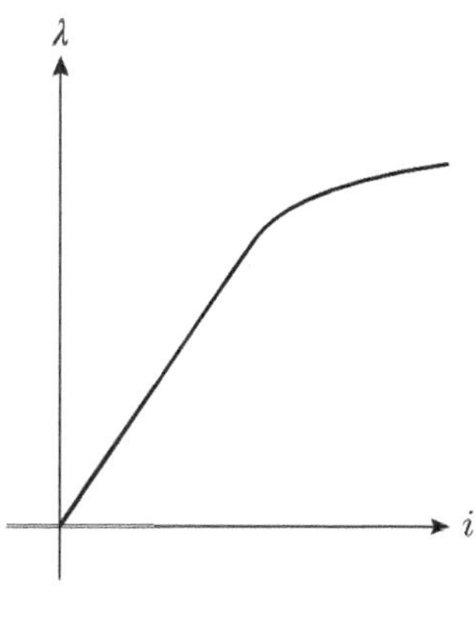

to the saturation of the ferromagnetic member. We will assume that the λ versus i characteristic is a single-valued function in that there is only one value of λ for a given value of i, whereupon hysteresis is neglected. When a ferromagnetic material is subjected to a magnetic field, its magnetic characteristics tend to aid in producing flux. This continues at a near-linear rate until this rate begins to decrease causing the knee of the magnetization curve. As the strength of the applied field continues to increase (by increasing i), the rate will continue to decrease and ultimately the magnetic characteristic of the ferromagnetic material approaches that of air. It is clear that after the knee of the magnetization curve the field energy W_f is no longer equal to the coenergy W_c.

Example 1B ***Inductance calculations of a magnetic circuit***

The parameters of the magnetic circuit shown in Figure 1.1 are: $r = 1\ \Omega$, $N = 100$ turns, $v = 10$ V, $\ell_i = 40$ cm, $x = 3$ mm, $A_i = A_g = 40$ cm^2, $\mu_0 = 4\pi \times 10^{-7}$ H/m, $\mu_{ri} = 1000$. Determine (a) the steady-state total flux Φ, assume $\mathcal{R}_l = 6.77 \times 10^7\ \text{H}^{-1}$; (b) the leakage inductance L_l, the magnetizing inductance L_m, and the self-inductance L; (c) express the energy from the source to the system following a step applied voltage.

(a) It would be helpful to refer to Figure 1.2. Therein,

$$\mathcal{R}_i = \frac{\ell_i}{\mu_{ri}\mu_0 A_i}$$

$$= \frac{40 \times 10^{-2}}{(1000)(4\pi \times 10^{-7})(4 \times 10^{-4})} = 7.95 \times 10^5\ \text{H}^{-1} \tag{1B.1}$$

$$\mathcal{R}_g = \frac{x}{\mu_{rg}\mu_0 A_g}$$

$$= \frac{3 \times 10^{-3}}{(1)(4\pi \times 10^{-7})(4 \times 10^{-4})} = 5.97 \times 10^6\ \text{H}^{-1} \tag{1B.2}$$

$$\Phi_m = \frac{Ni}{\mathcal{R}_i + \mathcal{R}_g}$$

$$= \frac{100\left(\frac{10}{1}\right)}{7.95 \times 10^5 + 5.97 \times 10^6} = 1.48 \times 10^{-4}\ \text{Wb} \tag{1B.3}$$

$$\Phi_l = \frac{Ni}{\mathcal{R}_l}$$

$$= \frac{100\left(\frac{10}{1}\right)}{6.77 \times 10^7} = 1.48 \times 10^{-5}\ \text{Wb} \tag{1B.4}$$

$$\Phi = \Phi_l + \Phi_m$$

$$= 1.48 \times 10^{-5} + 1.48 \times 10^{-4} = 1.63 \times 10^{-4}\ \text{Wb} \tag{1B.5}$$

(b) $L_l = \dfrac{N^2}{\mathcal{R}_l}$

$$= \frac{(100)^2}{6.77 \times 10^7} = 0.148\ \text{mH} \tag{1B.6}$$

$$L_m = \frac{N^2}{\mathcal{R}_m}$$

$$= \frac{(100)^2}{(7.96 \times 10^5) + (5.97 \times 10^5)} = 1.48\ \text{mH} \tag{1B.7}$$

$$L = L_l + L_m$$

$$= (0.148 \times 10^{-3}) + (1.48 \times 10^{-3}) = 1.63\ \text{mH} \tag{1B.8}$$

(c) The expression for the total energy from the source, W_E, is

$$W_E = \int vi\, dt \tag{1B.9}$$

From (1.32), which can now be written

$$v = ri + L\frac{di}{dt} \tag{1B.10}$$

Substituting (1B.10) into (1B.9) yields

$$\begin{aligned} W_E &= \int ri^2 dt + \int L\frac{di}{dt} i\, dt \\ &= r\int i^2 dt + \int L\, i\, di \\ &= r\int i^2 dt + \frac{1}{2}Li^2 \\ &= r\int i^2 dt + \frac{1}{2}(L_l + L_m)i^2 \end{aligned} \tag{1B.11}$$

The first term on the right side of (1B.11) is the energy dissipated in the resistor, while the second term is the energy stored in the inductance (magnetic field).

Since the expression for the current after the step voltage is applied is

$$i = \frac{v}{r}(1 - e^{-t/\tau}) \tag{1B.12}$$

where τ is the time constant L/r, the energy supplied by the source may be expressed

$$\begin{aligned} W_E &= r\int \left[\frac{v}{r}(1 - e^{-t/\tau})\right]^2 dt + \frac{1}{2}Li^2 \\ &= r\left(\frac{v}{r}\right)^2 \int (1 - 2e^{-t/\tau} + e^{-2t/\tau})\, dt + \frac{1}{2}Li^2 \\ &= \frac{v^2}{r}\left[t + 2\tau e^{-t/\tau} - \frac{\tau}{2}e^{-2t/\tau}\right] + \frac{1}{2}L\left(\frac{v}{r}\right)^2 (1 - e^{-t/\tau})^2 \end{aligned} \tag{1B.13}$$

For $t \gg \tau$,

$$W_E \cong \frac{v^2}{r}t + \frac{1}{2}L\left(\frac{v}{r}\right)^2 \quad (1B.14)$$

It is interesting to note that the energy stored in the inductance converges to a constant; however, the resistor continues to dissipate energy at a rate of v^2/r.

SP1.3.1 Determine two changes in the magnetic circuit shown in Figure 1.1 that would result in interchanging the position of the poles. [Reverse the direction of positive current or reverse the sense of the winding]

SP1.3.2 Express (a) W_f and (b) W_c if $L = 1$ H and the current is $I = I_p \cos \omega_e t$. [(a) $\frac{1}{2}I_p^2\left(\frac{1}{2} + \frac{1}{2}\cos 2\omega_e t\right)$, (b) same]

SP1.3.3 Re-draw the λ versus i plot in Figure 1.3. Indicate the change in the slope as the length of the air-gap increases. [slope decreases]

SP1.3.4 Consider Figure 1.1. Assume the reluctance of the iron may be neglected. Express the energy stored in the air gap in terms of its air-gap mmf (mmf_{ag}) and its reluctance $\mathcal{R}_g$. [$\frac{(\text{mmf}_{ag})^2}{2\mathcal{R}_g}$]

1.4 Stationary Coupled Circuits – The Transformer

Faraday's law tells us that a voltage is induced in an electric circuit due to a change of flux linkages. In the case of stationary circuits, such as the transformer, a change of flux linkages occurs due to time-varying currents. Therefore, in an ac power system, we can change the level of the voltage and current. That is, power can be generated at convenient voltage and current levels and transmitted at a high voltage with a low current to minimize transmission losses and then lower the voltage to a safe level at the point of usage.

Flux linkages of an electric circuit can also be changed, and thus voltages induced, due to relative motion between magnetic fields. This allows power to be generated at the source where mechanical motion is converted to electric power (generator action). At the load, electric power is often converted back to mechanical motion (motor action). The work of Faraday and later the inventions and development of the ac power system by Tesla and Westinghouse doomed Edison's dc power system in the very early twentieth century. We'll start our analytical journey into the world of Faraday and Tesla with the transformer.

1.4.1 Magnetically Linear Transformer

Two coupled circuits are shown in Figure 1.5. Actually, all we have done is "closed up" the air gap of Figure 1.1 and added a second coil. About all we have to concern ourselves with is how to handle the coupling between coils since in the previous section we became familiar with leakage flux and magnetizing flux. The flux linkages are

$$\lambda_1 = N_1\Phi_1 \tag{1.55}$$

$$\lambda_2 = N_2\Phi_2 \tag{1.56}$$

where Φ_1 and Φ_2 are the total flux linking N_1 and N_2, respectively. From Figure 1.5

$$\begin{aligned}\Phi_1 &= \Phi_{l1} + \Phi_{m1} + \Phi_{m2}\\ &= \frac{N_1 i_1}{\mathcal{R}_{l1}} + \frac{N_1 i_1}{\mathcal{R}_m} + \frac{N_2 i_2}{\mathcal{R}_m}\end{aligned} \tag{1.57}$$

$$\begin{aligned}\Phi_2 &= \Phi_{l2} + \Phi_{m2} + \Phi_{m1}\\ &= \frac{N_2 i_2}{\mathcal{R}_{l2}} + \frac{N_2 i_2}{\mathcal{R}_m} + \frac{N_1 i_1}{\mathcal{R}_m}\end{aligned} \tag{1.58}$$

Mutual coupling occurs since Φ_{m2}, which is the flux created by current flowing in coil N_2, links coil N_1. Likewise, Φ_{m1}, which is the flux created by i_1 current flowing in coil N_1, links coil N_2.

Substituting (1.57) into (1.55) and (1.58) into (1.56) yields

$$\lambda_1 = \frac{N_1^2}{\mathcal{R}_{l1}} i_1 + \frac{N_1^2}{\mathcal{R}_m} i_1 + \frac{N_1 N_2}{\mathcal{R}_m} i_2 \tag{1.59}$$

$$\lambda_2 = \frac{N_2^2}{\mathcal{R}_{l2}} i_2 + \frac{N_2^2}{\mathcal{R}_m} i_2 + \frac{N_2 N_1}{\mathcal{R}_m} i_1 \tag{1.60}$$

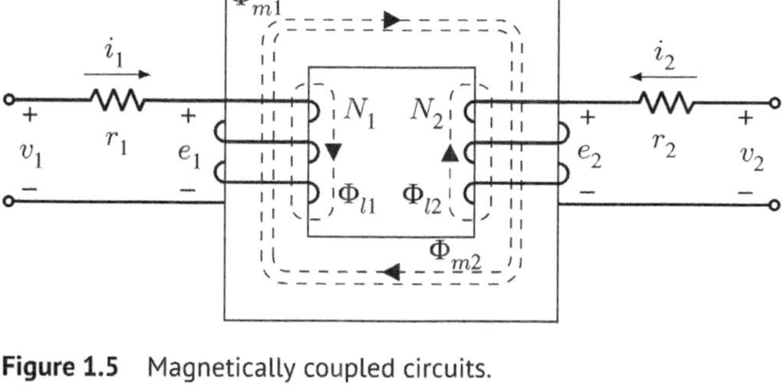

Figure 1.5 Magnetically coupled circuits.

The coefficients of the first two terms of (1.59) and (1.60) are the self-inductances L_{11} and L_{22}, respectively, that is

$$L_{11} = \frac{N_1^2}{\mathcal{R}_{l1}} + \frac{N_1^2}{\mathcal{R}_m} = L_{l1} + L_{m1} \tag{1.61}$$

$$L_{22} = \frac{N_2^2}{\mathcal{R}_{l2}} + \frac{N_2^2}{\mathcal{R}_m} = L_{l2} + L_{m2} \tag{1.62}$$

The coefficient of the last term of (1.59) and (1.60) is the mutual inductance

$$L_{12} = L_{21} = \frac{N_1 N_2}{\mathcal{R}_m} \tag{1.63}$$

We can now write λ_1 and λ_2 as

$$\lambda_1 = L_{11} i_1 + L_{12} i_2 \tag{1.64}$$

$$\lambda_2 = L_{22} i_2 + L_{21} i_1 \tag{1.65}$$

It is important to note that L_{m1}, L_{m2}, L_{12}, and L_{21} have a common term $\mathcal{R}_m$ which is the reluctance of the mutual path around the ferromagnetic core. This fact will allow us to derive a very useful equivalent circuit; however, before getting started on that, let us go back to (1.59) and (1.60). The mutual inductance can be positive or negative in these equations depending upon the relative directions of Φ_{m1} and Φ_{m2}. If the assumed positive directions of the currents or the sense of the winding of the coil causes Φ_{m1} and Φ_{m2} to aid each other, L_{12} (L_{21}) would be positive as in the case here. If they oppose L_{12} (L_{21}) would be negative.

Now, λ_1 may be written

$$\begin{aligned} \lambda_1 &= L_{l1} i_1 + L_{m1} i_1 + L_{12} i_2 \\ &= L_{l1} i_1 + L_{m1} \left(i_1 + \frac{N_2}{N_1} i_2 \right) \end{aligned} \tag{1.66}$$

Here we multiplied L_{12}, given by (1.63), by $\frac{N_1}{N_2}$ whereupon

$$\begin{aligned} \frac{N_1}{N_2} L_{12} &= \left(\frac{N_1}{N_2} \right) \frac{N_1 N_2}{\mathcal{R}_m} \\ &= L_{m1} \end{aligned} \tag{1.67}$$

thus,

$$L_{12} = \frac{N_2}{N_1} L_{m1} \tag{1.68}$$

From (1.66), we see there is a common inductance L_{m1} that carries the currents i_1 and $\frac{N_2}{N_1} i_2$. We have referred $\frac{N_2}{N_1} i_2$ to the winding with N_1 turns; if we would have expressed λ_2 rather than λ_1 we could have referred i_1 to the winding with N_2 turns by $\frac{N_1}{N_2}i_1$. Let's stay with $\frac{N_2}{N_1}i_2$, that is generally written, for compactness, as

$$i_2' = \frac{N_2}{N_1}i_2 \tag{1.69}$$

or

$$N_1 i_2' = N_2 i_2 \tag{1.70}$$

From (1.70), i_2' flowing in N_1 produces the same mmf as i_2 flowing in N_2. To maintain the same power in primed variables

$$v_2' i_2' = v_2 i_2 \tag{1.71}$$

Thus

$$v_2' = \frac{N_1}{N_2}v_2 \tag{1.72}$$

Since λ_2 has units of V s, the same turns ratio used for voltages can be applied to it; thus,

$$\lambda_2' = \frac{N_1}{N_2}\lambda_2 \tag{1.73}$$

There is considerable algebraic manipulation to get to the equivalent circuit. We'll only set forth the steps. First, let us write the voltage equations in terms of 1 and 2 variables.

$$v_1 = r_1 i_1 + \frac{d\lambda_1}{dt} \tag{1.74}$$

$$v_2 = r_2\, i_2 + \frac{d\lambda_2}{dt} \tag{1.75}$$

where r_1 and r_2 are the resistances of winding 1 and winding 2, respectively. Since we are referring to the two variables to the one winding, (1.74) remains unchanged in form; however, (1.75) becomes

$$v_2' = r_2' i_2' + \frac{d\lambda_2'}{dt} \tag{1.76}$$

The flux linkage equations become

$$\lambda_1 = L_{l1} i_1 + L_{m1}(i_1 + i_2') \tag{1.77}$$

$$\lambda_2' = L_{l2}' i_2' + L_{m1}(i_1 + i_2') \tag{1.78}$$

where

$$L_{l2}' = \left(\frac{N_1}{N_2}\right)^2 L_{l2} \tag{1.79}$$

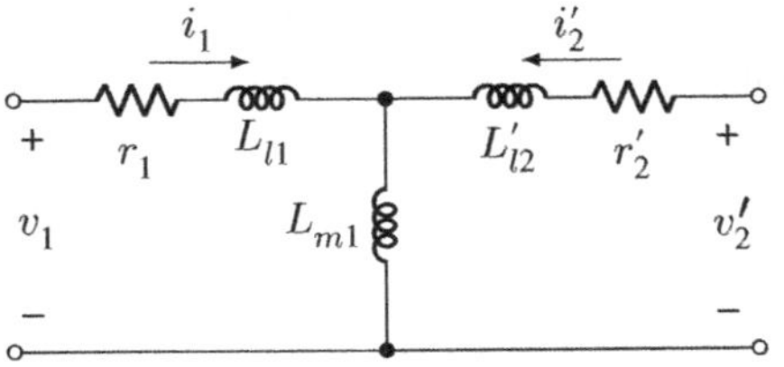

Figure 1.6 Transformer equivalent *T* circuit with winding 1 selected as reference winding.

$$r_2' = \left(\frac{N_1}{N_2}\right)^2 r_2 \tag{1.80}$$

The equivalent circuit is shown in Figure 1.6. This circuit is valid for all modes of operation, transient and steady state; however, it is valid only for a magnetically linear system. Also, we will find that with a slight modification the equivalent circuit shown in Figure 1.6 is valid for the steady-state performance of an induction machine.

1.4.2 Field Energy

Before leaving our work with transformers, let us derive an expression for the energy stored in the field W_f. In the case of a single coil, the field energy was $\frac{1}{2}Li^2$. Here, we have two coils with mutual coupling between them. In this case, the energy from the electrical system is

$$W_e = \int e_1 i_1 dt + e_2 i_2 dt \tag{1.81}$$

Since r_1 and r_2 are external to the respective coils and using unreferred variables

$$e_1 = \frac{d\lambda_1}{dt} \tag{1.82}$$

$$e_2 = \frac{d\lambda_2}{dt} \tag{1.83}$$

Substituting (1.82) and (1.83) into (1.81) yields

$$W_e = \int i_1 d\lambda_1 + i_2 d\lambda_2 \tag{1.84}$$

From (1.64) and (1.65)

$$d\lambda_1 = L_{11} di_1 + L_{12} di_2 \tag{1.85}$$

$$d\lambda_2 = L_{22} di_2 + L_{12} di_1 \tag{1.86}$$

Substituting (1.85) and (1.86) into (1.84) and since $W_f = W_e$, we have

$$W_f = \int [i_1(L_{11} di_1 + L_{12} di_2) + i_2(L_{22} di_2 + L_{12} di_1)] \tag{1.87}$$

For a magnetically linear system these integrals can be calculated in two steps; first let the current in winding 1 be the variable of integration and vary it from zero to i_1 while holding i_2 and di_2 at zero. Next hold the current in winding 1 at i_1 ($di_1 = 0$) and let the current in winding 2 be the variable of integration varying from zero to i_2. Performing the first step, we have

$$W_{f(1)} = \int_0^{i_1} L_{11}\xi d\xi = \frac{1}{2}L_{11}i_1^2 \tag{1.88}$$

For the second step

$$\begin{aligned} W_{f(2)} &= \int_0^{i_2} L_{12}i_1 d\xi + L_{22}\xi d\xi \\ &= L_{12}i_1i_2 + \frac{1}{2}L_{22}i_2^2 \end{aligned} \tag{1.89}$$

Thus

$$W_f = \frac{1}{2}L_{11}i_1^2 + L_{12}i_1i_2 + \frac{1}{2}L_{22}i_2^2 \tag{1.90}$$

The energy stored in the coupling field is W_f given by (1.90) less the energy stored in the leakage inductances. In a problem at the end of the chapter, you are asked to express (1.90) in terms of i_1 and i_2' and to identify the energy stored in the coupling field.

The pattern is clear, the field energy may be expressed for multiple electrical inputs (coils) as $\frac{1}{2}$ the self-inductance of each coil times the square of the current flowing in the coil and the mutual inductance between each set of coils times the product of the current flowing in the mutually coupled coils. For three coupled coils

$$W_f = \frac{1}{2}L_{11}i_1^2 + \frac{1}{2}L_{22}i_2^2 + \frac{1}{2}L_{33}i_3^2 + L_{12}i_1i_2 + L_{13}i_1i_3 + L_{23}i_2i_3 \tag{1.91}$$

Now, for multiple electrical inputs (windings), (1.54) becomes

$$\sum_{j=1}^{J} \lambda_j i_j = W_f + W_c \tag{1.92}$$

where J is the number of windings. Thus,

$$W_c = \sum_{j=1}^{J} \lambda_j i_j - W_f \tag{1.93}$$

In SP1.4.3, you are asked to prove that W_c is W_f using (1.91) and (1.93).

Example 1C ***Parameters of the transformer equivalent circuit***
It is instructive to illustrate the method of deriving an equivalent T circuit from open- and short-circuit measurements of the transformer. When winding 2 of the

two-winding transformer shown in Figure 1.6 is open circuited and a voltage of 110 V (rms) at 60 Hz is applied to winding 1, the average power supplied to winding 1 is 6.66 W. The measured current in winding 1 is 1.05 A (rms). Next, with winding 2 short-circuited, the current flowing in winding 1 is 2 A when the applied voltage is 30 V at 60 Hz. The average input power is 44 W. If we assume $L_{l1} = L'_{l2}$, an approximate equivalent T circuit can be determined from these measurements with winding 1 selected as the reference winding.

The average power supplied to winding 1 may be expressed from (1.31) as

$$P_1 = |\widetilde{V}_1|\,|\widetilde{I}_1| \cos \varphi_{pf} \tag{1C.1}$$

where

$$\varphi_{pf} = \theta_{ev}(0) - \theta_{ei}(0) \tag{1C.2}$$

Here, $\widetilde{V}_1$ and $\widetilde{I}_1$ are phasors with the positive direction of $\widetilde{I}_1$ taken in the direction of voltage drop, and $\theta_{ev}(0)$ and $\theta_{ei}(0)$ are the phase angles of $\widetilde{V}_1$ and $\widetilde{I}_1$, respectively. Solving for φ_{pf} during the open-circuit test, we have

$$\varphi_{pf} = \cos^{-1}\frac{P_1}{|\widetilde{V}_1|\,|\widetilde{I}_1|} = \cos^{-1}\frac{6.66}{(110)(1.05)} = 86.7^\circ \tag{1C.3}$$

Although $\varphi_{pf} = -86.7^\circ$ is also a legitimate solution of (1C.3), the positive value is taken since $\widetilde{V}_1$ leads $\widetilde{I}_1$ in an inductive circuit. With winding 2 open-circuited, the input impedance of winding 1 is

$$Z = \frac{\widetilde{V}_1}{\widetilde{I}_1} = r_1 + j(X_{l1} + X_{m1}) \tag{1C.4}$$

With $\widetilde{V}_1$ as the reference phasor, $\widetilde{V}_1 = 110\underline{/0^\circ}$, $\widetilde{I}_1 = 1.05\underline{/-86.7^\circ}$. Thus,

$$r_1 + j(X_{l1} + X_{m1}) = \frac{110\underline{/0^\circ}}{1.05\underline{/-86.7^\circ}} = 6 + j104.6\,\Omega \tag{1C.5}$$

From (1C.5), $r_1 = 6\,\Omega$. We also see from (1C.5) that $X_{l1} + X_{m1} = 104.6\,\Omega$.

For the short-circuit test, we will assume that $\widetilde{I}_1 = -\widetilde{I}'_2$ since transformers are designed so that at rated frequency $X_{m1} \gg |r'_2 + jX'_{l2}|$. Hence, using (1C.1) again,

$$\varphi_{pf} = \cos^{-1}\frac{44}{(30)(2)} = 42.8^\circ \tag{1C.6}$$

In this case, the input impedance is $Z = (r_1 + r'_2) + j(X_{l1} + X'_{l2})$. This may be determined as

$$Z = \frac{30\underline{/0^\circ}}{2\underline{/-42.8^\circ}} = 11 + j10.2\,\Omega \tag{1C.7}$$

Hence, $r'_2 = 11 - r_1 = 5\,\Omega$ and, since it is assumed that $X_{l1} = X'_{l2}$, both are $10.2/2 = 5.1\,\Omega$. Therefore, $X_{m1} = 104.6 - 5.1 = 99.5\,\Omega$. In summary, $r_1 = 6\,\Omega$,

$L_{l1} = 13.5$ mH, $L_{m1} = 263.9$ mH, $r'_2 = 5\,\Omega$, $L'_{l2} = 13.5$ mH. It is left to the reader to verify the conversion from X's to L's.

SP1.4.1 Consider the transformer and parameters calculated in Example 1C. Winding 2 is short-circuited and 12 V (dc) is applied to winding 1. Calculate the steady-state values of i_1 and i_2. Repeat with winding 2 open-circuited. [$I_1 = 2$ A and $I_2 = 0$ in both cases]

SP1.4.2 Calculate the no-load current (winding 2 open-circuited) for the transformer given in Example 1C if $V_1 = \sqrt{2}10\cos 100t$. [$\tilde{I}_1 = 0.352\underline{/-77.8^\circ}$ A]

SP1.4.3 Determine W_c from (1.91) and (1.93). Show that $W_c = W_f$.

SP1.4.4 Use the equivalent T circuit given in Figure 1.6 to identify the energy stored in the coupling field. [$\frac{1}{2}L_{m1}(i_1^2 + i'^2_2) + L_{m1}i_1i'_2$]

1.5 Two- and Three-phase Systems

High-voltage transmission, most inverter-supplied electric drives, and the alternator of your car are examples of three-phase systems. Although two-phase systems are not common, a two-phase system is far less involved when it comes to machine analysis than its three-phase counterpart. Fortunately, once the derivations have been set forth for a two-phase machine, the extension to a three-phase machine is straightforward and easily achieved. This section is devoted to the introduction of two- and three-phase systems.

By definition, a two-phase set of variables is balanced if the variables are equal-amplitude sinusoidal quantities in time quadrature (90° out of time phase). A three-phase set of variables is balanced if the sinusoidal variables are equal-amplitude quantities that are 120° out of time phase with each other.

1.5.1 Two-phase Systems

In the broadest sense of the above definition, two-phase balanced sets may be expressed as

$$f_a(t) = \pm f(t)\cos\theta_{ef} \tag{1.94}$$

$$f_b(t) = \pm f(t)\sin\theta_{ef} \tag{1.95}$$

where

$$\theta_{ef}(t) = \int_0^t \omega_e(\xi)d\xi + \theta_{ef}(0) \tag{1.96}$$

In (1.94) and (1.95), $f(t)$ can represent voltage, current, or flux linkage. The amplitude may be any function of time; however, for steady-state balanced conditions $f(t)$ is a constant. In (1.96), ω_e is the electrical angular velocity and ξ is a dummy variable of integration. Equations (1.94) and (1.95) express four balanced two-phase sets. Like signs of (1.94) and (1.95) define balanced sets where $f_a(t)$ leads $f_b(t)$ by 90°: an *ab* sequence; for unlike signs $f_a(t)$ lags $f_b(t)$ by 90°: a *ba* sequence.

For steady-state balanced conditions ω_e is constant and (1.96) becomes

$$\theta_{ef}(t) = \omega_e t + \theta_{ef}(0) \tag{1.97}$$

Whereupon (1.94) and (1.95) are written as

$$F_a(t) = \pm\sqrt{2}F\cos[\omega_e t + \theta_{ef}(0)] \tag{1.98}$$

$$F_b(t) = \pm\sqrt{2}F\sin[\omega_e t + \theta_{ef}(0)] \tag{1.99}$$

For like signs of (1.98) and (1.99) $\tilde{F}_a = j\tilde{F}_b$; for unlike signs $\tilde{F}_a = -j\tilde{F}_b$.

Most large horsepower electric machines are three phase and smaller household machines are single phase. Therefore, it can be argued that there is no need to treat two-phase devices. This is an understandable position; however, we will find that it is convenient to first analyze a two-phase device and then extend this to a three-phase device. This turns out to be much less involved and more instructive than to start the analysis with the three-phase device. Although this is perhaps not evident at this juncture, it will become very evident as we get into the later chapters.

It is helpful to take a brief look at the stator winding arrangement of the two-phase machine, shown in Figure 1.7. The windings are assumed to be identical in parameters and distribution. The displacement around the stator is denoted ϕ_s. The two windings are displaced 90° from each other. The voltage equations may be written (dropping the functional notation)

$$v_{as} = r_s i_{as} + \frac{d\lambda_{as}}{dt} \tag{1.100}$$

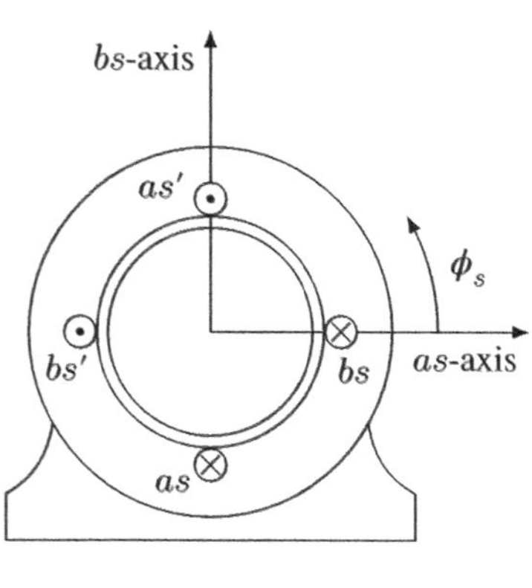

Figure 1.7 Elementary two-pole two-phase concentrated stator windings.

$$v_{bs} = r_s i_{bs} + \frac{d\lambda_{bs}}{dt} \tag{1.101}$$

In matrix form

$$\mathbf{v}_{abs} = \mathbf{r}_s \mathbf{i}_{abs} + p\boldsymbol{\lambda}_{abs} \tag{1.102}$$

The stator windings are identical, and the air gap is uniform. The magnetic axes are orthogonal, (thus $L_{asbs} = 0$) and the flux linkage equations may be written

$$\begin{aligned}\lambda_{as} &= L_{asas} i_{as} \\ &= (L_{ls} + L_{ms}) i_{as}\end{aligned} \tag{1.103}$$

$$\begin{aligned}\lambda_{bs} &= L_{bsbs} i_{bs} \\ &= (L_{ls} + L_{ms}) i_{bs}\end{aligned} \tag{1.104}$$

where L_{ls} is the leakage inductance and L_{ms} is the magnetizing inductance of the stator windings. In matrix form

$$\begin{bmatrix} \lambda_{as} \\ \lambda_{bs} \end{bmatrix} = \begin{bmatrix} L_{ss} & 0 \\ 0 & L_{ss} \end{bmatrix} \begin{bmatrix} i_{as} \\ i_{bs} \end{bmatrix} \tag{1.105}$$

or

$$\boldsymbol{\lambda}_{abs} = \mathbf{L}_s \mathbf{i}_{abs} \tag{1.106}$$

where

$$L_s = \begin{bmatrix} L_{ss} & 0 \\ 0 & L_{ss} \end{bmatrix} \tag{1.107}$$

and

$$L_{ss} = L_{ls} + L_{ms} \tag{1.108}$$

An important feature of multiphase systems is that the instantaneous power is constant for balanced operation. You are asked to show this in SP1.5.2. Recall that in a single-phase system the instantaneous power has an average value and a double frequency component.

1.5.2 Three-phase Systems

A three-phase balanced set may be expressed as

$$f_a(t) = f(t) \cos \theta_{ef} \tag{1.109}$$

$$f_b(t) = f(t) \cos \left(\theta_{ef} - \frac{2}{3}\pi \right) \tag{1.110}$$

$$f_c(t) = f(t) \cos \left(\theta_{ef} + \frac{2}{3}\pi \right) \tag{1.111}$$

This set is referred to as an *abc* sequence, since $f_a(t)$ leads $f_b(t)$ by 120° and $f_b(t)$ leads $f_c(t)$ by 120°. An *acb* sequence is

$$f_a(t) = f(t)\cos\theta_{ef} \tag{1.112}$$

$$f_b(t) = f(t)\cos\left(\theta_{ef} + \frac{2}{3}\pi\right) \tag{1.113}$$

$$f_c(t) = f(t)\cos\left(\theta_{ef} - \frac{2}{3}\pi\right) \tag{1.114}$$

For steady-state balanced conditions, the *abc* sequence may be written

$$F_a(t) = \sqrt{2}F\cos\left[\omega_e t + \theta_{ef}(0)\right] \tag{1.115}$$

$$F_b(t) = \sqrt{2}F\cos\left[\omega_e t - \frac{2}{3}\pi + \theta_{ef}(0)\right] \tag{1.116}$$

$$F_c(t) = \sqrt{2}F\cos\left[\omega_e t + \frac{2}{3}\pi + \theta_{ef}(0)\right] \tag{1.117}$$

With $\tilde{F}_a = F\underline{/\theta_{ef}(0)}$, $\tilde{F}_b = F\underline{/\theta_{ef}(0) - \frac{2}{3}\pi}$, and $\tilde{F}_c = F\underline{/\theta_{ef}(0) + \frac{2}{3}\pi}$. For an *acb* sequence $\tilde{F}_b$ and $\tilde{F}_c$ are interchanged.

The windings of a three-phase stator are shown in Figure 1.8. The magnetic axes of the windings are displaced 120° and the windings are often "wye-connected"

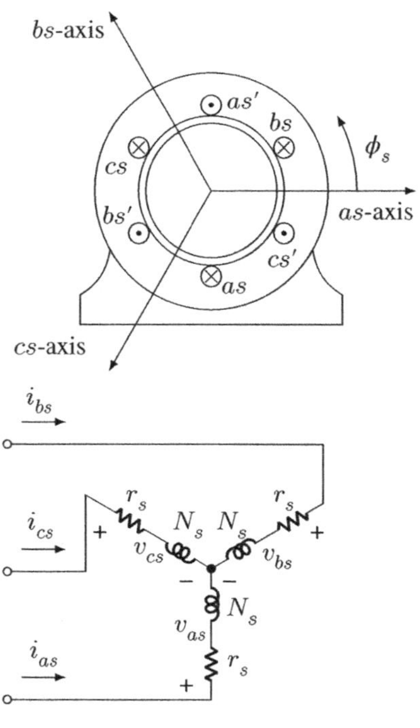

Figure 1.8 Elementary two-pole three-phase concentrated stator windings.

as shown. We'll talk more about three-phase connections later. The voltage equations may be written (again dropping the functional notation)

$$v_{as} = r_s i_{as} + \frac{d\lambda_{as}}{dt} \tag{1.118}$$

$$v_{bs} = r_s i_{bs} + \frac{d\lambda_{bs}}{dt} \tag{1.119}$$

$$v_{cs} = r_s i_{cs} + \frac{d\lambda_{cs}}{dt} \tag{1.120}$$

In matrix form

$$\mathbf{v}_{abcs} = \mathbf{r}_s \mathbf{i}_{abcs} + p\boldsymbol{\lambda}_{abcs} \tag{1.121}$$

Since the windings are displaced 120° from each other, there is a mutual coupling between the stator windings. Let us assume that we can move the *bs* winding clockwise through the iron until it is "on top" of the *as* winding at $\phi_s = 0$. The coupling would be maximum positive. Now assume we can rotate the *bs* winding counterclockwise back to $\phi_s = 120°$ where the mutual inductance between the *as* and *bs* windings can be approximated as

$$\begin{aligned} L_{asbs} &= L_{ms} \cos 120° \\ &= -\frac{1}{2} L_{ms} \end{aligned} \tag{1.122}$$

where L_{ms} is the magnetizing inductance of the stator windings. Following this same approach, we can express the flux-linkage matrix as

$$\begin{bmatrix} \lambda_{as} \\ \lambda_{bs} \\ \lambda_{cs} \end{bmatrix} = \begin{bmatrix} L_{ss} & -\frac{1}{2}L_{ms} & -\frac{1}{2}L_{ms} \\ -\frac{1}{2}L_{ms} & L_{ss} & -\frac{1}{2}L_{ms} \\ -\frac{1}{2}L_{ms} & -\frac{1}{2}L_{ms} & L_{ss} \end{bmatrix} \begin{bmatrix} i_{as} \\ i_{bs} \\ i_{cs} \end{bmatrix} \tag{1.123}$$

where

$$L_{ss} = L_{ls} + L_{ms} \tag{1.124}$$

Equation (1.123) may also be written as

$$\boldsymbol{\lambda}_{abcs} = \mathbf{L}_s \mathbf{i}_{abcs} \tag{1.125}$$

Example 1D ***Voltage equations for a three-wire system***

A three-phase stator similar to that given in Figure 1.8 is connected to a three-phase source as shown in Figure 1D.1. Assume the stator is symmetrical, that is, the windings have the same resistance and same number of turns and displaced 120°. The stator could be that of induction, synchronous, or brushless dc machines. The source voltages e_{ga}, e_{gb}, and e_{gc} may be of any form. Express v_{as}, v_{bs}, and v_{cs} in terms of e_{ga}, e_{gb}, and e_{gc}.

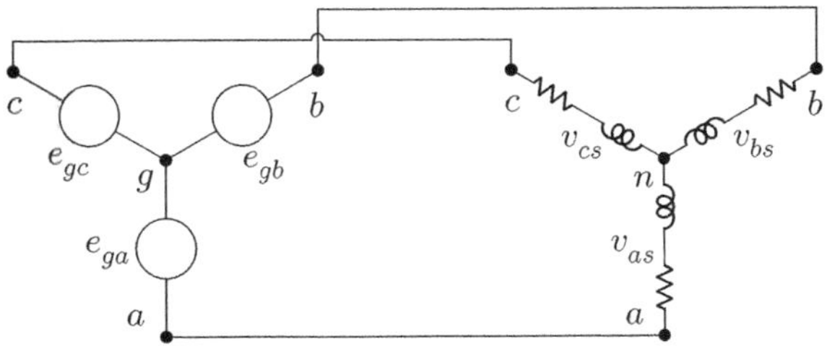

Figure 1D.1 Three-phase source connected to symmetrical stator windings.

From Figure 1D.1 we can write

$$e_{ga} = v_{as} + v_{ng} \tag{1D.1}$$

$$e_{gb} = v_{bs} + v_{ng} \tag{1D.2}$$

$$e_{gc} = v_{cs} + v_{ng} \tag{1D.3}$$

Adding (1D.1)–(1D.3) yields

$$e_{ga} + e_{gb} + e_{gc} = v_{as} + v_{bs} + v_{cs} + 3v_{ng} \tag{1D.4}$$

Let us look at $v_{as} + v_{bs} + v_{cs}$. From (1.118)–(1.120)

$$v_{as} + v_{bs} + v_{cs} = r_s(i_{as} + i_{bs} + i_{cs}) + p(\lambda_{as} + \lambda_{bs} + \lambda_{cs}) \tag{1D.5}$$

In a three-wire, wye-connected stator the sum of $i_{as} + i_{bs} + i_{cs}$ must be zero regardless of the form of the currents. Now from (1.123)

$$\lambda_{as} + \lambda_{bs} + \lambda_{cs} = L_{ss}(i_{as} + i_{bs} + i_{cs}) - L_{ms}(i_{as} + i_{bs} + i_{cs}) = 0 \tag{1D.6}$$

Thus,

$$v_{as} + v_{bs} + v_{cs} = 0 \tag{1D.7}$$

The question is, will this be the case when we bring the rotor into play? We will find that for the electromechanical devices we'll consider, it will be true. Substituting (1D.7) into (1D.4) yields

$$v_{ng} = \frac{1}{3}(e_{ga} + e_{gb} + e_{gc}) \tag{1D.8}$$

Going back to (1D.1)–(1D.3) we can write

$$\begin{aligned} v_{as} &= e_{ga} - v_{ng} \\ &= \frac{2}{3}e_{ga} - \frac{1}{3}(e_{gb} + e_{gc}) \end{aligned} \tag{1D.9}$$

$$\begin{aligned} v_{bs} &= e_{gb} - v_{ng} \\ &= \frac{2}{3}e_{gb} - \frac{1}{3}(e_{gc} + e_{ga}) \end{aligned} \tag{1D.10}$$

$$v_{cs} = e_{gc} - v_{ng} = \frac{2}{3}e_{gc} - \frac{1}{3}(e_{ga} + e_{gb}) \quad (1D.11)$$

We will make use of these equations when considering electric drives.

SP1.5.1 In Figure 1D.1, let $e_{ga} = 1$, $e_{gb} = 0$, $e_{gc} = \cos\omega_e t$. Determine v_{as}, v_{bs}, and v_{cs}. $[\frac{2}{3} - \frac{1}{3}\cos\omega_e t; -\frac{1}{3} - \frac{1}{3}\cos\omega_e t; -\frac{1}{3} + \frac{2}{3}\cos\omega_e t]$. Note that $v_{as} + v_{bs} + v_{cs} = 0$.

SP1.5.2 In a two-phase system, let $V_a = \sqrt{2}V\cos[\omega_e t + \theta_{ev}(0)]$, $I_a = \sqrt{2}I \cos[\omega_e t + \theta_{ei}(0)]$, $V_b = \sqrt{2}V\sin[\omega_e t + \theta_{ev}(0)]$, and $I_b = \sqrt{2}I\sin[\omega_e t + \theta_{ei}(0)]$. Show that the total instantaneous power is $P = 2VI\cos[\,\theta_{ev}(0) - \theta_{ei}(0)]$.

1.6 Problems

1.1 Using the same circuit parameters as given in Example 1A; $R = 6\ \Omega$, $L = 20$ mH, and $C = 1 \times 10^3\ \mu$F, consider the circuit shown in Figure 1.9 Let the 60-Hz source voltage be $\tilde{V}_a = 110\underline{/0°}$. Determine all phasors and draw the phasor diagram.

1.2 For the magnetic system in Example 1B, let $V(t) = \sqrt{2}\ 10\cos 377t$. Establish $I(t)$ and express $W_f(t)$.

1.3 During the short-circuit test of the transformer in Example 1C, which is a step-down distribution transformer, it was noticed that the I_2 was 6 A. In redoing the open-circuit test the open-circuit current $|\tilde{I}_1|$ was not 1.05 A but 0.6 A. Upon checking the wiring it was discovered that the lower terminal of N_1 was mistakenly connected to the top terminal of N_2 (see Figure 1.5) and the 110-V source was connected between the top terminal of N_1 and the

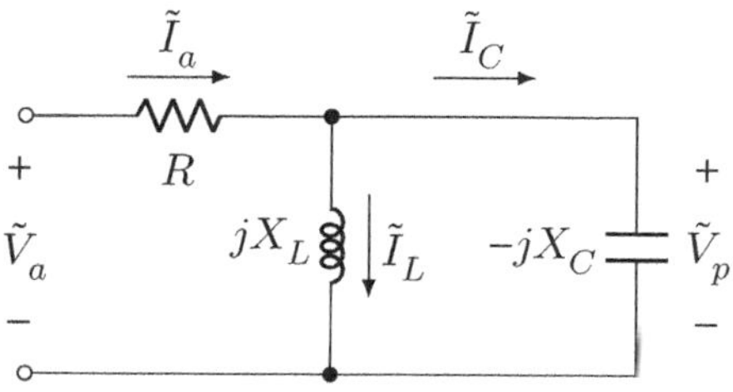

Figure 1.9 Series-parallel circuit.

bottom terminal of N_2. Justify that $|\tilde{I}_1| = 0.6$ A. The equivalent T circuit will not yield the correct answer.

1.4 During the open-circuit test performed in Example 1C the rms voltage across the open-circuit 2 winding was 34.8 V. Determine X_{m2} ($\omega_e L_{m2}$).

1.5 Express the field energy W_f stored in the fields of the coils shown in Figure 1.5 in terms of L_{l1}, L'_{l2}, L_{m1}, i_1, and i'_2. Identify the energy stored in the coupling field.

1.6 Consider Example 1D and Figure 1D.1. The load is symmetrical and the source voltages are given in Figure 1.10. (a) Plot v_{ng}, v_{as}, v_{bs}, and v_{cs}. (b) Connect n to g in Figure 1D.1 and repeat part (a).

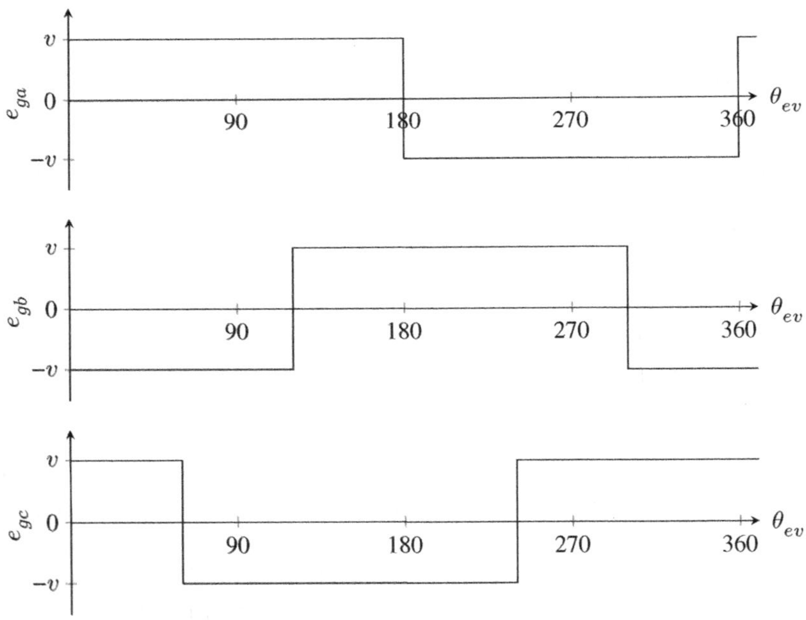

Figure 1.10 Waveforms of the source voltages of Figure 1D.1.

2

Fundamentals of Electric Machine Analysis

2.1 Introduction

It is told that Nikola Tesla, as a young student in Graz, Austria, tried to convince his professor that there must be a better electric machine design than the direct current (dc) motor being demonstrated to the class. Tesla argued that the arcing of the commutator was very undesirable and inefficient. The commutator, along with a carbon-brush arrangement, mechanically switched the direction of the currents flowing in the rotor circuits to create a stationary magnet field, orthogonal with the magnetic field created by direct current flowing in the stator winding. The professor chided young Tesla and told the class there was no other way to design an electric motor. It took Tesla several years to develop the concept of the rotating magnetic field which is also referred to as the rotating air-gap magnetomotive force (mmf) and/or rotating poles [1]. This led to his invention of alternating current (ac) machines and multiphase power transmission, and he and Westinghouse used this concept to light the 1893 World Columbian Exposition in Chicago. This marked the birth of the ac power system as we know it today and the beginning of the demise of the dc power system being promoted by Edison.

In 1929, Robert H. Park published a change of variables [2] that revolutionized the analysis of ac machines and ac power systems. It was recently found that this change of variables or transformation is readily obtained from the expression of Tesla's rotating magnetic field [3]. Although Park's transformation was originally used in power system analysis, it also set the stage for the analysis of electric drive systems which began to evolve with the advent of the controlled rectifier in the 1950s. Although Tesla and Park are deceased, their work will live as long as there are ac power systems and electric drives.

In this chapter, Tesla's rotating magnet field and the change of variables are introduced. This is achieved by starting with a two-phase system, which allows these concepts to be set forth without the complication of tedious trigonometry.

Introduction to the Analysis of Electromechanical Systems, First Edition.
Paul C. Krause, Oleg Wasynczuk, and Timothy O'Connell.

Once this is done, the extension to a three-phase machine/system is straightforward and direct. This is a logical approach since for balanced conditions the expression for Tesla's rotating magnetic field (air-gap mmf) differs only by a constant when going from a two-phase to a three-phase machine.

These basic concepts establish a foundation for the analysis of electric machines, power systems, and electric drives that are now being used in, among other things, ground vehicles, aircraft, spacecraft, and industrial applications. Once this foundation is established, synchronous and induction machines are analyzed. This choice of machines allows us to set the stage for the analysis of the brushless dc motor and the field-oriented induction motor drives as well as the synchronous generator, which is the backbone of the ac power system.

2.2 Coupled Circuits in Relative Motion

In the previous chapter, we considered the "transformer" coupling that exists between two stationary windings. An induced voltage due to a time-varying current (transformer action) is only "half" of Faraday's Law. In the case of electric machines, there can be windings on the stationary member (stator) and windings on the rotating member (rotor). Transformer coupling may exist between only stator windings or between only rotor windings. Transformer coupling may also exist between the stator and rotor windings depending upon their relative positions. In addition, there is a voltage induced in the circuits due to the rate of change of the relative position of the stator and rotor windings, commonly referred to as a speed voltage. In this section, we will take our first look at the magnetic coupling when windings are in relative motion and what Faraday's law tells us about the voltages that can be induced.

The rotational device shown in Figure 2.1 will be used to illustrate windings in relative motion. This device consists of two coils each containing several turns of a conductor. Winding 1 has N_1 turns and it is on the stationary member (stator); winding 2 has N_2 turns and it is on the rotating member (rotor). The ⊗ indicates that the assumed direction of positive current flowing in the conductors is into the paper, where ⊙ indicates positive current flowing in the conductors is out of the paper. In a practical device, the turns of a winding are distributed over an arc (often 30–60°) of the stator or rotor; however, in this introductory consideration, it is sufficient to assume, for now, that the turns are concentrated in one position, as shown in Figure 2.1. Also, the length of the air gap between the stator and rotor is shown exaggerated relative to the actual inside diameter of the stator.

The voltage equations may be written as

$$v_1 = r_1 i_1 + \frac{d\lambda_1}{dt} \tag{2.1}$$

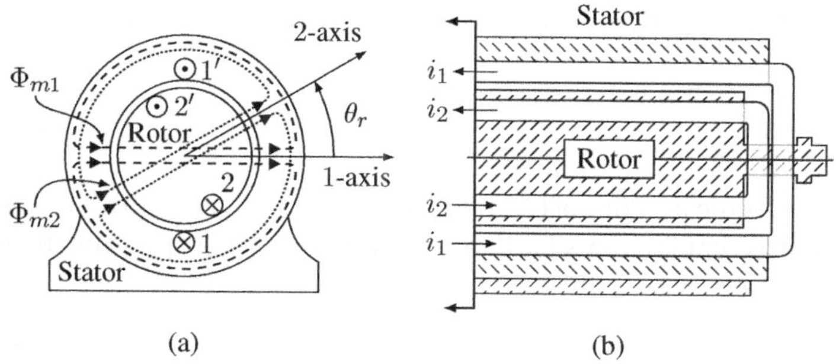

Figure 2.1 Elementary rotational electromechanical device. (a) End view; (b) cross-sectional view.

$$v_2 = r_2 i_2 + \frac{d\lambda_2}{dt} \tag{2.2}$$

where r_1 and r_2 are the resistances of windings 1 and 2, respectively. The magnetic system is assumed to be linear; therefore, the flux linkages may be expressed as

$$\lambda_1 = L_{11} i_1 + L_{12} i_2 \tag{2.3}$$

$$\lambda_2 = L_{21} i_1 + L_{22} i_2 \tag{2.4}$$

Since the air gap is uniform, the self-inductances L_{11} and L_{22} are constants and may be expressed in terms of leakage and magnetizing inductances as

$$\begin{aligned} L_{11} &= L_{l1} + L_{m1} \\ &= \frac{N_1^2}{\mathcal{R}_{l1}} + \frac{N_1^2}{\mathcal{R}_m} \end{aligned} \tag{2.5}$$

$$\begin{aligned} L_{22} &= L_{l2} + L_{m2} \\ &= \frac{N_2^2}{\mathcal{R}_{l2}} + \frac{N_2^2}{\mathcal{R}_m} \end{aligned} \tag{2.6}$$

where $\mathcal{R}_m$ is the reluctance of the complete magnetic path of Φ_{m1} and Φ_{m2}, which is through the rotor and stator iron and twice across the air gap $(\mathcal{R}_i + 2\mathcal{R}_g)$. Clearly, $\mathcal{R}_m$ is the same for the magnetic system established by either winding 1 or 2.

Take a moment to note the designation of 1 axis and 2 axis in Figure 2.1. These axes denote the positive direction of the respective magnetic systems with the assumed positive direction of current flowing in the windings (right-hand rule). Now let us consider L_{12}. (Is it clear that $L_{12} = L_{21}$?) When θ_r is zero, the coupling between windings 1 and 2 is maximum. In particular, with $\theta_r = 0$ the magnetic system of winding 1 aids that of winding 2 with positive currents assumed. Hence, the mutual inductance is positive and

$$L_{12}(0) = \frac{N_1 N_2}{\mathcal{R}_m} \tag{2.7}$$

When $\theta_r = \frac{1}{2}\pi$, the windings are orthogonal. The mutual coupling is zero. Hence,

$$L_{12}\left(\frac{1}{2}\pi\right) = 0 \tag{2.8}$$

When $\theta_r = \pi$, $L_{12}(\pi)$ is the negative of (2.7) and at $\theta_r = \frac{3}{2}\pi$, L_{12} is zero; the same as (2.8). Let us make it as simple as possible by assuming that as a first approximation the mutual inductance may be written

$$L_{12}(\theta_r) = L_{sr}\cos\theta_r \tag{2.9}$$

where L_{sr} is the amplitude of the sinusoidal mutual inductance between the stator and rotor windings as given by (2.7).

The flux linkages are given by (2.3) and (2.4), let's rewrite these equations indicating the functionality

$$\lambda_1(i_1, i_2, \theta_r) = L_{11}i_1 + L_{12}(\theta_r)i_2 \tag{2.10}$$

$$\lambda_2(i_1, i_2, \theta_r) = L_{12}(\theta_r)i_1 + L_{22}i_2 \tag{2.11}$$

The total derivative of λ_1 becomes

$$\begin{aligned}\frac{d\lambda_1}{dt} &= \frac{\partial(L_{11}i_1)}{\partial i_1}\frac{di_1}{dt} + \frac{\partial(L_{11}i_1)}{\partial i_2}\frac{di_2}{dt} + \frac{\partial(L_{11}i_1)}{\partial \theta_r}\frac{d\theta_r}{dt} \\ &\quad + \frac{\partial(L_{12}i_2)}{\partial i_1}\frac{di_1}{dt} + \frac{\partial(L_{12}i_2)}{\partial i_2}\frac{di_2}{dt} + \frac{\partial(L_{12}i_2)}{\partial \theta_r}\frac{d\theta_r}{dt}\end{aligned} \tag{2.12}$$

Only the first, fifth, and sixth terms on the right-hand side of (2.12) are nonzero; thus, v_1 can be written

$$v_1 = r_1 i_1 + L_{11}\frac{di_1}{dt} + L_{sr}\cos\theta_r\frac{di_2}{dt} - i_2 L_{sr}\sin\theta_r\frac{d\theta_r}{dt} \tag{2.13}$$

Following the same procedure, we can express v_2 as

$$v_2 = r_2 i_2 + L_{22}\frac{di_2}{dt} + L_{sr}\cos\theta_r\frac{di_1}{dt} - i_1 L_{sr}\sin\theta_r\frac{d\theta_r}{dt} \tag{2.14}$$

Faraday's law has given us the three terms on the right-hand side of (2.13) and (2.14). The first of these terms is due to the change of current flowing in the winding; this term would exist independent of the other winding. The second term is due to the "transformer" coupling from one winding to the other. The degree and sense of the coupling would depend upon the relative position or displacement between the windings. The last of these three terms is the voltage induced due to the rate of change of the displacement between windings; commonly referred to as the "speed voltage."

2.2.1 Field Energy

Regardless if the electromechanical device is magnetically linear or nonlinear, the λi relationship is a single-valued function with the assumptions we have made. Thus, the energy stored in the field is a single-valued state function, meaning that for a given i_1, i_2, and θ_r (or x in case of translational motion) there is only one value of W_f. Therefore, even though (1.87) was derived for magnetically linear, stationary coupled circuits it is also valid for magnetically linear coupled circuits in relative motion. Thus, from (1.90)

$$W_f = \frac{1}{2}L_{11}i_1^2 + i_1 i_2 L_{sr}\cos\theta_r + \frac{1}{2}L_{22}i_2^2 \tag{2.15}$$

The assumption of a magnetically linear system may not be as limiting as it might first appear. In many cases, most of the mmf is dropped across the air gap of the electromechanical device, and therefore most of the field energy W_f is stored in the air gap and air is a magnetically linear, conservative medium.

Example 2A ***Inductances of windings in relative motion***
Consider the windings in relative motion shown in Figure 2A.1. The outer stator housing has been omitted from that shown in Figure 2.1. We'll often omit this housing for convenience. Winding 3 has been added to the rotor. Windings 2 and 3 are identical and displaced by $\frac{\pi}{2}$ relative to each other. Express L_{11}, L_{22}, L_{33}, L_{12}, L_{13}, L_{23}, and v_1.

We have L_{11} and L_{22} from (2.5) and (2.6), respectively. It follows that

$$L_{33} = \frac{N_3^2}{\mathcal{R}_{l3}} + \frac{N_3^2}{\mathcal{R}_m} \tag{2A.1}$$

Since windings 2 and 3 are identical $N_2 = N_3$ thus $L_{22} = L_{33}$. In the case of L_{12} and L_{13}

$$L_{12} = -\frac{N_1 N_2}{\mathcal{R}_m}\sin\theta_r \tag{2A.2}$$

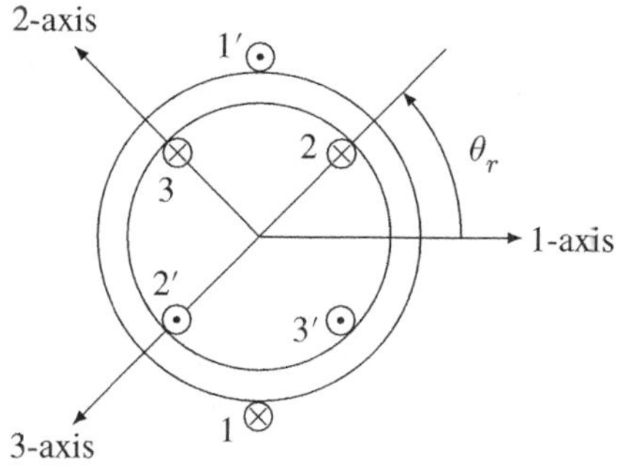

Figure 2A.1 Windings in relative motion (stator housing omitted).

$$L_{13} = -\frac{N_1 N_3}{\mathcal{R}_m}\cos\theta_r \tag{2A.3}$$

Since windings 2 and 3 are orthogonal, $L_{23} = 0$. The voltage v_1 may be written by using (2.13) or (2.14) as a guide.

$$v_1 = r_1 i_1 + L_{11}\frac{di_1}{dt} - L_{sr}\sin\theta_r\frac{di_2}{dt} - L_{sr}\cos\theta_r\frac{di_3}{dt}$$
$$- i_2 L_{sr}\cos\theta_r\frac{d\theta_r}{dt} + i_3 L_{sr}\sin\theta_r\frac{d\theta_r}{dt} \tag{2A.4}$$

where

$$L_{sr} = \frac{N_1 N_2}{\mathcal{R}_m} \tag{2A.5}$$

and $N_2 = N_3$.

SP2.2.1 Consider Figure 2.1 $I_1 = 1$ A, $L_{sr} = 0.1$ H, $L_{11} = 10L_{sr}$, $\omega_r = 100$ rad/s, $\theta_r(0) = 0$, and winding 2 is open circuited. Express V_2. [$V_2 = -10\sin 100t$]

SP2.2.2 Calculate the field energy for SP2.2.1. [0.5 J]

SP2.2.3 Repeat SP2.2.1 if $I_1 = 1\cos 100t$. [$V_2 = -10\sin 200t$]

2.3 Electromagnetic Force and Torque

Our goal in this section is to derive an expression for the electromagnetic torque developed in rotational systems or force in translational systems. Although there are several approaches that could be used for this derivation, with the background that we have established the "energy balance" approach is perhaps most convenient [4]. This method is quite direct and results in an easy-to-use expression for evaluating torque or force in magnetically linear electromagnetic systems.

Electromechanical devices consist of an electrical system, a coupling field, and a mechanical system. The electric and the mechanical systems can either supply or absorb energy by way of the coupling field which can either be electromagnetic, which will be our focus, or electrostatic. The coupling field enables the electrical and mechanical systems to interact thereby providing a means of transferring energy between the two systems while at the same time providing a means of storing energy. A block diagram depicting this interaction and possible directions of energy flow is given in Figure 2.2.

The energy balance is shown in Figure 2.3. W_E and W_M are positive for energy being supplied to the electromechanical device from an external electrical system

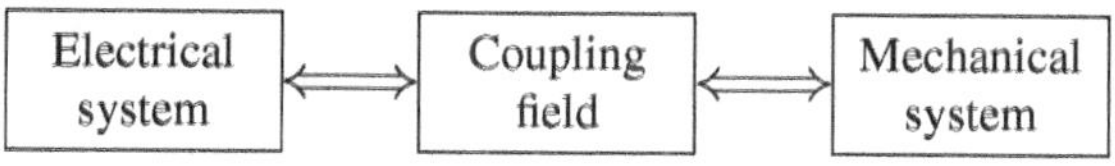

Figure 2.2 Block diagram of possible energy interchange in an elementary electromechanical system.

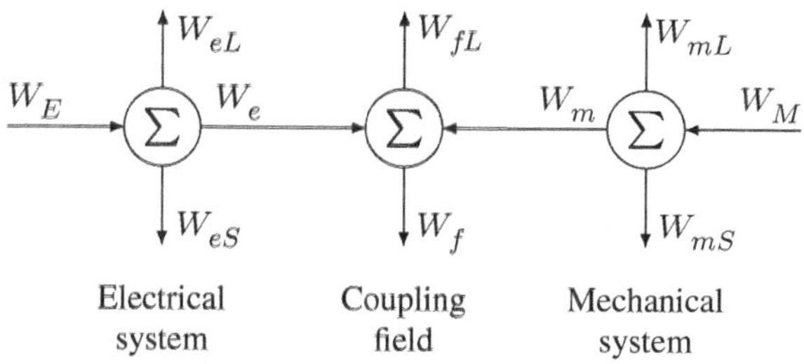

Figure 2.3 Energy balance.

(W_E) and mechanical system (W_M). Likewise W_e and W_m are positive when supplying energy to the coupling field in the electrical form (W_e) and mechanical form (W_m).

In Figure 2.3, the subscripts E and e pertain to the electrical system; M and m to the mechanical; and f to the field. Also, the subscripts S and L indicate energy stored and energy lost, respectively. The energy W_{eL} is the energy lost due to i^2r; W_{eS} is the energy stored in a field external to the coupling field. Some authors consider energy stored in the field of the leakage inductances as W_{eS} while others will include the leakage flux in the coupling field. This is irrelevant since the energy stored in the field of the leakage inductance is generally not a function of the mechanical motion of the electromechanical device.

The energy W_{mL} represents energy lost due to mechanical friction and windage losses. W_{mS} is the energy stored as kinetic energy in the rotating mass (rotor) or as potential energy in a spring.

The energy W_e is the energy coming into the coupling field from the electric system. The energy W_m is the energy coming into the coupling field from the mechanical system. The energy W_{fL} is the energy lost due to hysteresis loss and circulating currents induced in the mechanical components of the coupling field. Electric machines are designed to minimize these losses making the λi relationship approach a single-valued function as shown in Figure 1.4. We will neglect W_{fL}; whereupon, we can express the energy balance of the coupling field as

$$W_f = W_e + W_m \tag{2.16}$$

where W_f is the energy stored in the coupling field [5].

Figure 2.1a is shown again in Figure 2.4 wherein the flux streamlines and the outside housing have been omitted from the cross-sectional view as was done in Figure 2A.1. Also electromagnetic torque T_e, the load torque T_L, and the angular

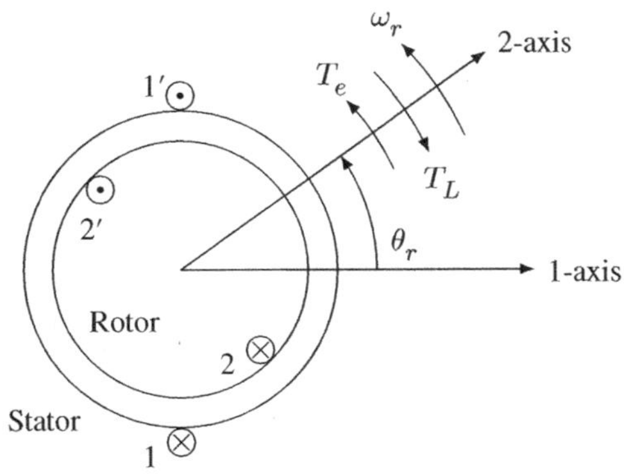

Figure 2.4 Elementary rotational electromechanical device.

velocity of the rotor ω_r are indicated. Note that T_e is positive in the direction of positive angular displacement of the rotor θ_r, while the load torque T_L is positive in the direction of negative θ_r.

The torque relationship between the electromechanical device and the mechanical system may be written

$$T_e = J\frac{d\omega_r}{dt} + B_m\omega_r + T_L \tag{2.17}$$

which is the rotational form of Newton's second law. Here, J is the inertia of the rotor and any tightly mechanically coupled rotating mass, B_m is the mechanical damping which is generally small and often neglected. Note that positive T_e acts to increase ω_r, whereas a positive load torque T_L would act to retard ω_r. Therefore, with T_e positive for motor action, then positive W_m entering the coupling field as shown in Figure 2.3 and as expressed in (2.16), would be

$$W_m = -\int T_e d\theta_r \tag{2.18}$$

The energy entering the coupling field from the two electrical sources shown in Figure 2.4 W_e is

$$W_e = \int e_1 i_1 dt + e_2 i_2 dt \tag{2.19}$$

where

$$e_1 = \frac{d\lambda_1}{dt} \tag{2.20}$$

$$e_2 = \frac{d\lambda_2}{dt} \tag{2.21}$$

and e_1 and e_2 are $v_1 - r_1 i_1$ and $v_2 - r_2 i_2$, respectively. Substituting (2.20) and (2.21) into (2.19) yields

$$W_e = \int i_1 d\lambda_1 + i_2 d\lambda_2 \tag{2.22}$$

For J electrical inputs (2.22) may be expressed

$$W_e = \int \sum_{j=1}^{J} i_j d\lambda_j \tag{2.23}$$

where j is an index $(1, 2, 3, \ldots, J)$ not to be confused with the imaginary part of a complex number or with inertia or Joules. Thus, (2.16) may now be written

$$W_f = \int \sum_{j=1}^{J} i_j d\lambda_j - \int T_e d\theta_r \tag{2.24}$$

In differential form, which we will use extensively, (2.24) can be written

$$T_e d\theta_r = \sum_{j=1}^{J} i_j d\lambda_j - dW_f \tag{2.25}$$

Equation (1.54) may be written for multiple electrical inputs as

$$\sum_{j=1}^{J} \lambda_j i_j = W_c + W_f \tag{2.26}$$

where W_c is the coenergy. If we take the derivative of (2.26), we have

$$\sum_{j=1}^{J} \lambda_j di_j + \sum_{j=1}^{J} i_j d\lambda_j = dW_c + dW_f \tag{2.27}$$

Solving (2.27) for $\sum_{j=1}^{J} i_j d\lambda_j$ and substituting the result into (2.25) yields

$$T_e d\theta_r = -\sum_{j=1}^{J} \lambda_j di_j + dW_c \tag{2.28}$$

Although (2.25) and (2.28) are valid for magnetically linear and nonlinear systems, we will consider only magnetically linear systems. Therefore, it is convenient to express the flux linkages in terms of currents and either θ_r for rotational systems or x for translational systems. Hence, we will choose currents and θ_r or x as independent variables. In this case, (2.28) is most convenient for obtaining an expression for T_e which can be written

$$T_e(\mathbf{i}, \theta_r) d\theta_r = -\sum_{j=1}^{J} \lambda_j(\mathbf{i}, \theta_r)\, di_j + dW_c(\mathbf{i}, \theta_r) \tag{2.29}$$

for compactness,

$$\mathbf{i} = (i_1, i_2, i_3, \ldots, i_J) \tag{2.30}$$

Equation (2.29) may now be written

$$T_e(\mathbf{i},\theta_r)d\theta_r = -\sum_{j=1}^{J}\lambda_j(\mathbf{i},\theta_r)di_j + \sum_{j=1}^{J}\frac{\partial W_c}{\partial i_j}(\mathbf{i},\theta_r)di_j + \frac{\partial W_c(\mathbf{i},\theta_r)}{\partial \theta_r}d\theta_r \tag{2.31}$$

Equating coefficients of $d\theta_r$

$$T_e(\mathbf{i},\theta_r) = \frac{\partial W_c}{\partial \theta_r} \tag{2.32}$$

In case of a translational system

$$f_e(\mathbf{i},x) = \frac{\partial W_c}{\partial x} \tag{2.33}$$

where f_e is the electromagnetic force and x is the displacement. Since we will consider only magnetically linear systems $W_c = W_f$.

Example 2B ***Force between pole faces of air gap***

Determine the force that exists between the pole faces of the elementary magnetic circuit given in Figure 1.1 and the parameters given in Example 1B. To make use of (2.33) to calculate the electromagnetic force f_e we must allow what is referred to as a virtual displacement. That is, we will assume that the air-gap length x increases in the positive direction by dx; therefore, according to (2.33) f_e will be positive if it acts to lengthen the air gap. If f_e is negative, it acts to shorten the air gap.

From (1.42), the self-inductance is

$$L = L_l + L_m \tag{2B.1}$$

where L_l is the leakage inductance and L_m is the magnetizing inductance. In particular,

$$L = \frac{N^2}{\mathcal{R}_l} + \frac{N^2}{\mathcal{R}_i + \mathcal{R}_g} \tag{2B.2}$$

The coenergy is

$$W_c = \frac{1}{2}(L_l + L_m)i^2 \tag{2B.3}$$

To determine f_e we must take the partial derivative of W_c with respect to x. Since L_l is not a function of x, we need only concern ourselves with L_m. We can write L_m from (1B.1) and (1B.2) as

$$\begin{aligned} L_m &= \frac{N^2}{\mathcal{R}_i + \mathcal{R}_g} \\ &= \frac{N^2}{\dfrac{\ell_i}{\mu_{ri}\mu_0 A_i} + \dfrac{x}{\mu_0 A_g}} \end{aligned} \tag{2B.4}$$

which can be written

$$L_m(x) = \frac{k_0}{k_1 + k_2 x} = \frac{\frac{k_0}{k_2}}{\frac{k_1}{k_2} + x} \tag{2B.5}$$

where

$$k_0 = N^2 = (1 \times 10^2)^2 = 1 \times 10^4 \text{ turns} \tag{2B.6}$$

$$k_1 = \frac{\ell_i}{\mu_{ri}\mu_0 A_i} = \frac{4 \times 10^{-1}}{(1 \times 10^3)(4\pi \times 10^{-7})(4 \times 10^{-4})} = 7.96 \times 10^5/\text{H} \tag{2B.7}$$

$$k_2 = \frac{1}{\mu_0 A_g} = \frac{1}{(4\pi \times 10^{-7})(4 \times 10^{-4})} = 1.99 \times 10^9 \text{ m/H} \tag{2B.8}$$

Therefore

$$W_c = \frac{\frac{1}{2}\frac{k_0}{k_2} i^2}{\frac{k_1}{k_2} + x} \tag{2B.9}$$

Now

$$f_e(i, x) = \frac{\partial W_c}{\partial x} = -\frac{\frac{1}{2}\left(\frac{k_0}{k_2} i^2\right)}{\left(\frac{k_1}{k_2} + x\right)^2} = -\frac{\frac{1}{2}(5.025 \times 10^{-6} \times 10^2)}{(3.995 \times 10^{-4} + 3 \times 10^{-3})^2} = -21.74 \text{ N} \quad \text{(an attractive force)} \tag{2B.10}$$

A force is established in a magnetic system to minimize the reluctance.

SP2.3.1 Neglect the reluctance of the ferromagnetic core in Example 2B and calculate f_e. [−27.92 N]

SP2.3.2 In a rotational system, the self-inductance of stator winding is $L_{11} = k_0 + k_1 \cos 2\theta_r$. There is only one stator winding and no rotor winding. Express T_e, which is referred to as the reluctance torque, if $I_1 = \cos \omega_e t$. $[-k_1 I_1^2 \sin 2\theta_r]$

SP2.3.3 Rearrange the magnetic circuit in Figure 1.1 to produce a repelling force between pole faces. [The force will always be in the direction to minimize the reluctance of the magnetic system]

2.4 Winding Configurations

The analysis of ac electromechanical motion devices is based upon the concept of the rotating air-gap mmf (rotating magnetic field) which was developed by Tesla in the late nineteenth century. Most ac electric machines are designed so that the stator windings produce a relatively good approximation of a sinusoidally distributed air-gap mmf, thereby minimizing the voltage and current harmonics. To establish a sinusoidal air-gap mmf waveform (often referred to as a *space sinusoid*), the physical distribution of the winding must approximate a sinusoidal space distribution. We haven't heard of many of these terms before; hopefully, this chapter will help us.

2.4.1 Concentrated Winding

Back in Section 2.2, we considered two concentrated windings in relative motion for a "first look" at the mutual inductance between stator and rotor windings. Our purpose now is to look a little further into the magnetic characteristics of the windings and how we are going to treat the windings in our analysis. Let us start by considering just a concentrated stator winding of N turns as shown in the cutaway view of the machine in Figure 2.5a and a cross-sectional view of the machine taken at some point along the length of the rotor shown in Figure 2.5b. In Figure 2.5a, r is the radius and l the axial length of the rotor. The stator winding consists of a coil of insulated wire of N turns in one slot. Current entering the bottom part of the coil is assumed to be positive, which is shown as ⊗ in Figure 2.5b and exiting the top part of the coil, shown as ⊙. It is important to realize that the length of the air gap is exaggerated relative to the rest of the machine; in fact, the radius of the inside circumference of the stator is essentially the radius of the rotor.

The streamlines of the flux established by current flowing in the stator winding are shown in Figure 2.5b. We see that flux leaves the stator where the winding exists and enters the air gap between $\frac{\pi}{2} < \phi_s < \frac{3}{2}\pi$, where ϕ_s is the position from the magnetic axis of the winding. This is the north pole of the stator, N^s, and it follows that S^s exits from $-\frac{\pi}{2} < \phi_s < \frac{\pi}{2}$.

Figure 2.5 Concentrated winding. (a) Cutaway view. (b) Cross-sectional view. (c) Equivalent magnetic circuit.

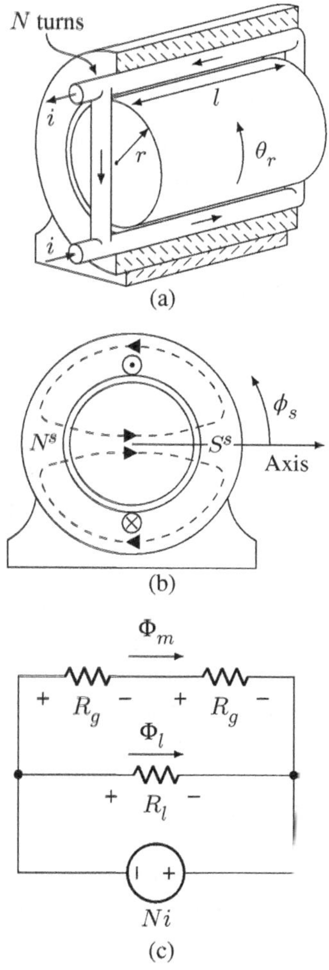

The equivalent magnetic circuit is shown in Figure 2.5c, where the reluctance of the iron has been neglected. The leakage flux, Φ_l, is the flux that fails to cross the air gap; instead occurring inside the stator slot itself and at the end turns of the coil. A detailed calculation of leakage flux is beyond the scope of this text. The reader may wish to refer to [6] for a comprehensive analysis. Since this flux path is mostly in air, Φ_l is small compared to Φ_m. Now, what about $\mathcal{R}_g$? Well, there are two air gaps and since the gaps are identical, one-half of the mmf is dropped across each air gap. That is the air-gap mmf of each air gap is $\frac{1}{2}(Ni)$. That's logical; however, we have yet to determine $\mathcal{R}_g$ for Figure 2.5b. Now, an acceptable approximation of the cross-sectional area of the flux path across each of the air gaps is

$$A_{ag} = \frac{1}{2}(2\pi rl) \tag{2.34}$$

where r is essentially the radius of the rotor and l is the axial length of the rotor. The reluctance is

$$\mathcal{R}_g = \frac{g}{\mu_{rg}\mu_0 \pi rl} \tag{2.35}$$

The flux Φ_m is

$$\begin{aligned}\Phi_m &= \frac{Ni}{2\mathcal{R}_g} \\ &= \frac{Ni\mu_{rg}\mu_0\pi rl}{2g}\end{aligned} \tag{2.36}$$

In (2.36) we have calculated the flux, Φ_m, as the total mmf drop across both air-gaps, which are in series, divided by the total reluctance. Clearly, we would get the same answer if we calculated Φ_m flowing across one air gap. In this case, the air-gap mmf is $\frac{1}{2}Ni$ and the reluctance is $\mathcal{R}_g$.

Before becoming too involved with winding distribution, let us talk a little more about air-gap mmf of an electric machine. For this purpose, we'll use Figure 2.6 which is a repeat of Figure 2.5b. The meter with an arrow indicator in Figure 2.6a is our newly invented mmfmeter, which we patterned after an old voltmeter. As in Figure 2.5b, the flux streamlines are shown and a positive (+) and a negative (−) meter deflection indicates the mmf drop across the air gap (air-gap mmf). It has a red probe (R) and a black probe (B). If we place the R-probe on the rotor side of the air gap and the B-probe on the stator side and if the constant current is positive ⊗ to ⊙, we would observe a positive deflection at $\phi_s = 0$. This is an mmf drop across the air gap from rotor to stator. If now we move the probes counterclockwise in unison around the air gap the meter reading remains essentially constant until we reach $\phi_s = 90°$. At $\phi_s = 90°$, it will decrease to zero since there is no air-gap mmf at $\phi_s = 90°$ and therefore flux does not cross the air gap. Moving the probes beyond $\phi_s = 90°$, the deflection will be negative and essentially equal in magnitude to the positive deflection we saw at $\phi_s = 0$. Ideally, the deflection will remain constant until we reach $\phi_s = 270°$. The deflection will then increase back to zero. Thus, for a positive coil current, a positive air-gap mmf exists from $-\frac{\pi}{2}\pi < \phi_s < \frac{\pi}{2}$ and negative from $\frac{\pi}{2} < \phi_s < \frac{3}{2}\pi$. Recall that this is only if the coil current is positive. We are going to let the current be sinusoidal later in this section. As we have mentioned, the flux, like current, must have a closed path; there is no point flux source.

The so-called developed winding is shown in Figure 2.6b,c. Hence, the machine is "cut" along the radius at $\phi_s = 0$ and the machine is "stretched out" with counterclockwise displacement about the stator viewed from right to left for positive ϕ_s

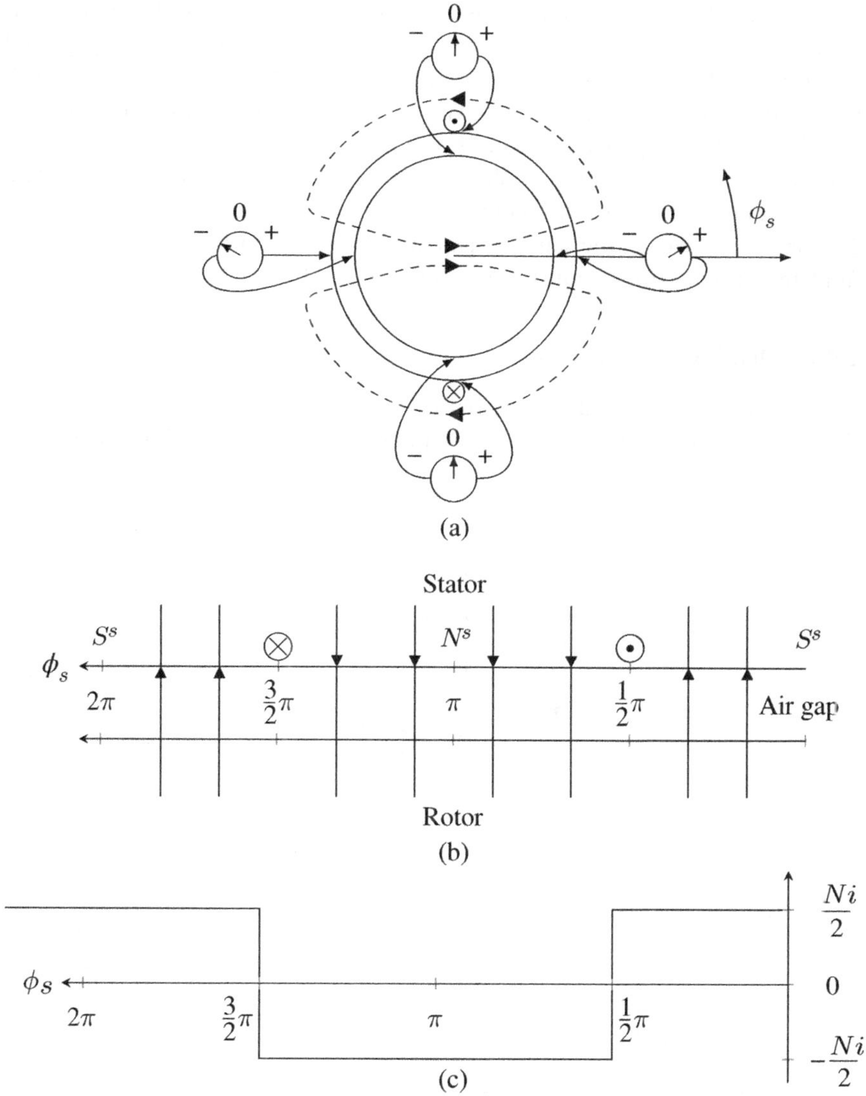

Figure 2.6 Air-gap mmf of concentrated stator winding. (a) "Measuring" the air-gap mmf. (b) Developed-winding view. (c) Plot of air-gap mmf versus ϕ_s.

position. This orientation allows us to place the stator above the rotor and maintain the rules of a magnetic system. The linear approximation of the circular air gap is allowed since the length of air gap is much less than the radius r of the rotor; that is $g \ll r$. Note the location of the stator poles in Figure 2.6b.

The plot of the air-gap mmf versus ϕ_s is shown in Figure 2.6c. Here, in accordance with established convention and assumptions made, positive air-gap mmf occurs when the mmf drop is from rotor to stator (as in Figure 2.6a) with the positive assumed current (flux) direction. We will make extensive use of the developed winding and the air-gap mmf versus ϕ_s plot.

We can get the same result as shown in Figure 2.6 by applying Ampere's law. Let us redraw Figure 2.6a in Figure 2.7. In Figure 2.7a path 1 is for $\phi_s = 0$, path 2 is for $\phi_s = \frac{\pi}{2}$. We get the same air-gap mmf as we did with the mmfmeter. The fundamental of Fourier's transform is shown as mmf_{as}.

2.4.2 Distributed Windings

Thus far, we have assumed that the stator windings are concentrated in one stator slot; however, the phase windings would normally be distributed over a 30°–60°

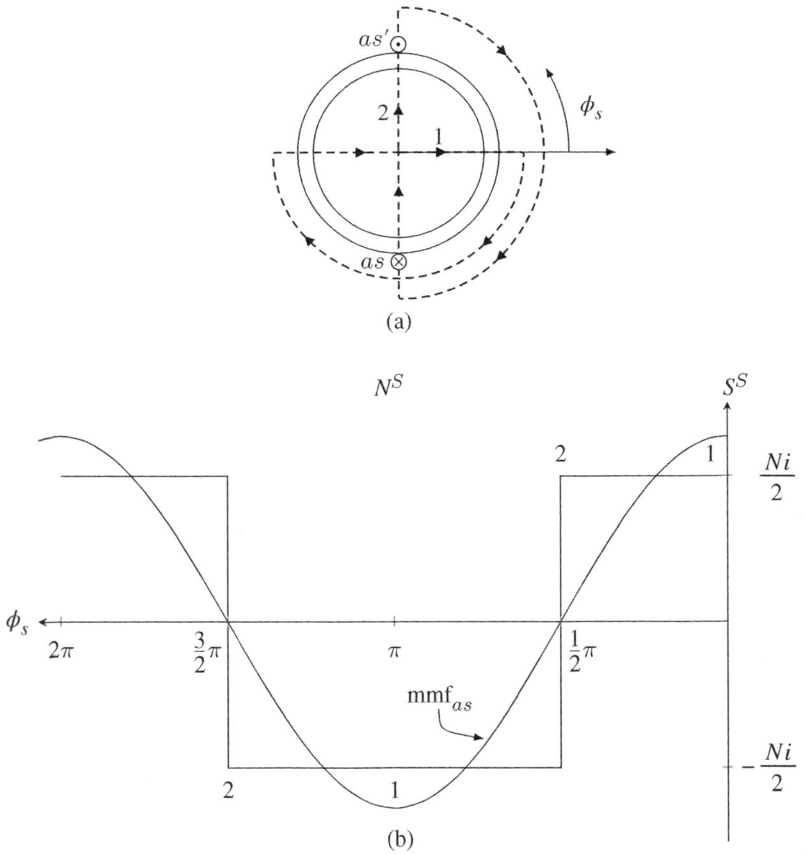

Figure 2.7 Applying Ampere's law to Figure 2.6a. (a) Ampere's law. (b) Developed winding and mmf plot.

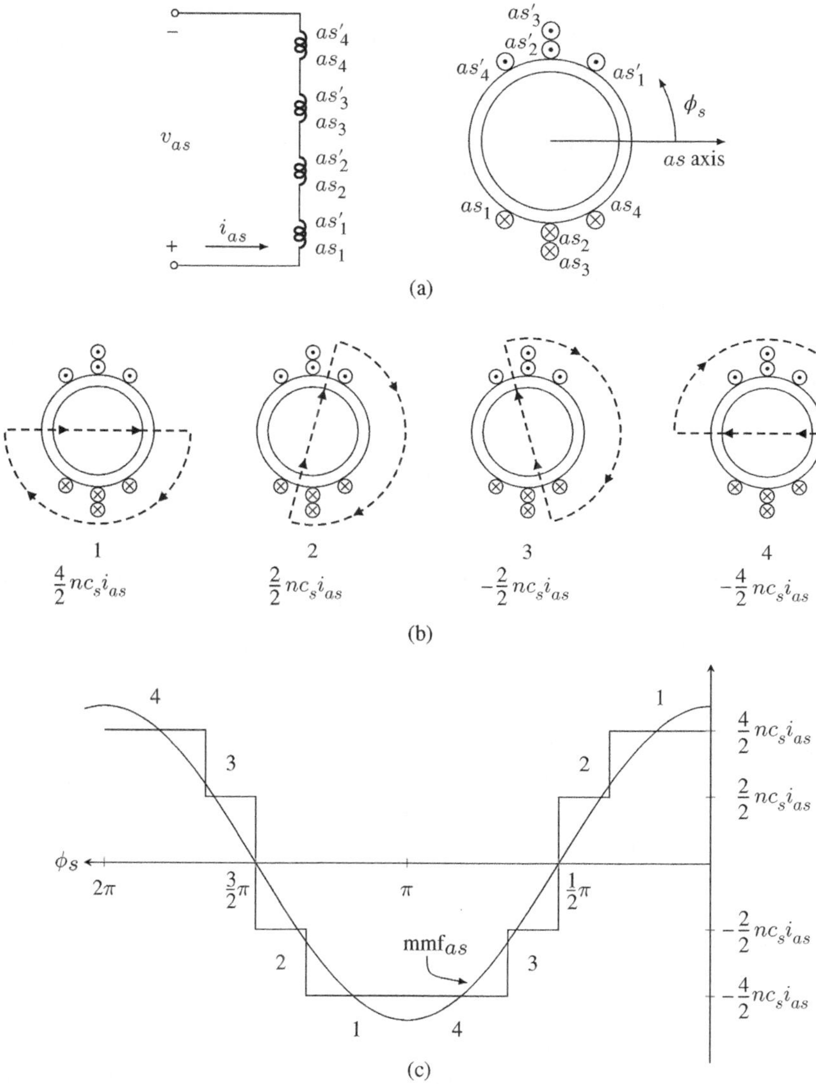

Figure 2.8 Elementary sinusoidally distributed winding. (a) Winding connections and distribution. (b) Ampere's law. (c) mmf_{as}.

arc. Let us consider Figure 2.8; here, we are calling the winding the *as* winding and we will assume that it is distributed in slots over the inner circumference of the stator, as shown in Figure 2.8.

The elementary sinusoidally distributed winding is depicted as a series of individual coils. Each coil has nc_s turns and is placed in a slot in the stator steel. If we follow the path of assumed positive current i_{as} flowing in the *as* winding, we see that current enters as_1, depicted by ⊗, to indicate that the assumed direction of positive current is down the length of the stator in an axial direction (into the paper). Current flows down the length of the stator, loops at the end, and flows back down the length of the stator and out at as_1', depicted by ⊙. Note that as_1 and as_1' are placed in stator slots that span π radians. This is referred to as the *winding pitch* of π radians, which is characteristic of a two-pole machine. We will talk more about poles later. Now, as_1 around to as_1' is referred to as a *coil* and as_1 or as_1' is a *coil side*. In practice, a coil will contain more than one conductor. Current flows into as_1 in a conductor and out as_1' via the same conductor. The conductor, which is insulated, may then be looped back to as_1 and the winding of the conductor around the $as_1 - as_1'$ path repeated, thereby forming a coil with numerous turns. The number of conductors in a coil side tells us the number of turns in the coil, which is denoted nc_s.

Once we have wound nc_s turns in the $as_1 - as_1'$ coil, we will take the same conductor and repeat this winding process to form the $as_2 - as_2'$ coil. We will assume that the same number of turns (nc_s) make up the $as_2 - as_2'$ coil as the $as_1 - as_1'$ coil and, similarly, for $as_3 - as_3'$ and $as_4 - as_4'$. We could have wound a different number of turns in each coil, but we will assume that this was not done. Once the winding is wound, we can use the right-hand rule to give a meaning to the *as* axis shown in Figure 2.8a. It is, by definition, the principal direction of the magnetic flux established by the assumed positive current flowing in the *as* winding. It is said to indicate the assumed positive direction of the magnetic axis of the *as* winding of this elementary sinusoidally distributed winding.

The paths of the line integrals, which span π radians, for Ampere's law are shown in Figure 2.8b and the resulting air-gap mmf is shown in Figure 2.8c where the sinusoidal approximation of the air-gap mmf is shown. This air-gap mmf approximation may be expressed

$$\text{mmf}_{as} = \frac{N_s}{2} i_{as} \cos \phi_s \tag{2.37}$$

where N_s is the amplitude of the fundamental component of the Fourier transform of the winding distribution. In Figure 2.8, $N_s = 2.37\ nc_s$. Note also that the sinusoidal distributed winding is denoted with ⊗ and ⊙ placed at the maximum winding density. We realize that this is the same way that we denoted a concentrated winding; however, from here on ⊗ and ⊙ will denote sinusoidally distributed winding.

The winding shown in Figure 2.8 is a very crude approximation of a sinusoidal distributed winding, one that would probably not be used except in inexpensive, low-power motors. In the case of a large generator, pains would be taken to distribute the winding much closer to a sinusoidal winding to minimize the voltage harmonics. Nevertheless, the applying of Ampere's law would be done the same.

Although we have introduced the sinusoidal distributed winding, we will not derive the mutual and self-inductances of sinusoidal distributed windings. In Chapter 1, we considered only concentrated windings and assumed that the mutual inductance between windings in relative motion varied sinusoidally. This, of course, was a first-order approximation; however, now that we have sinusoidally distributed windings, we will accept sinusoidal variations in the mutual inductances without proof.

Example 2C ***Air-gap mmf for a uniformly distributed winding***
Consider the uniformly distributed winding arrangement shown in Figure 2C.1. Each coil has nc_s turns and the current in each turn is i_{as} with the positive direction as shown. Follow the procedure used in Figure 2.8 to establish the air-gap mmf. Here $N_s = 1.534\, nc_s$.

SP2.4.1 Express the sinusoidal approximation of mmf_{as} if the assumed positive direction of i_{as} is reversed in Figure 2.8. [mmf_{as} = – Figure 2.8c]

SP2.4.2 Assume that winding 1 in Figure 2.1 has one turn. Sketch the air gap mmf due to current i_1 flowing in this winding. Let ϕ_s be the position from axis 1. [$\frac{1}{2}i_1$ for $-\frac{1}{2}\pi < \phi_s < \frac{1}{2}\pi$ and $-\frac{1}{2}i_1$ for $\frac{1}{2}\pi < \phi_s < \frac{2}{3}\pi$]

SP2.4.3 Repeat SP2.4.2 for winding 2. Let ϕ_r be the position from axis 2. [Same as SP2.4.2 with ϕ_s replaced by ϕ_r]

2.5 Rotating Air-gap mmf – Tesla's Rotating Magnetic Field

In the case of a single-phase stator winding, the sinusoidal approximation of the air-gap mmf is given by (2.37). Let's think about this for a minute; (2.37) is a product of i_{as} and $\cos\phi_s$; whereupon, the air-gap mmf will be nonexistent whenever $i_{as} = 0$. This is not too desirable since we will find that to produce a constant steady-state torque we need a constant amplitude rotating air-gap mmf (north and south poles) produced by the stator. Is it possible to place another winding on the stator and arrange its position and the time sequence of the current flowing in it so that its mmf would complement the mmf due the *as* winding resulting in a rotating sinusoidal air-gap mmf whose amplitude is constant? Yes, fortunately, and Tesla showed us how; the symmetrical two-pole two-phase stator is the most elementary example [1].

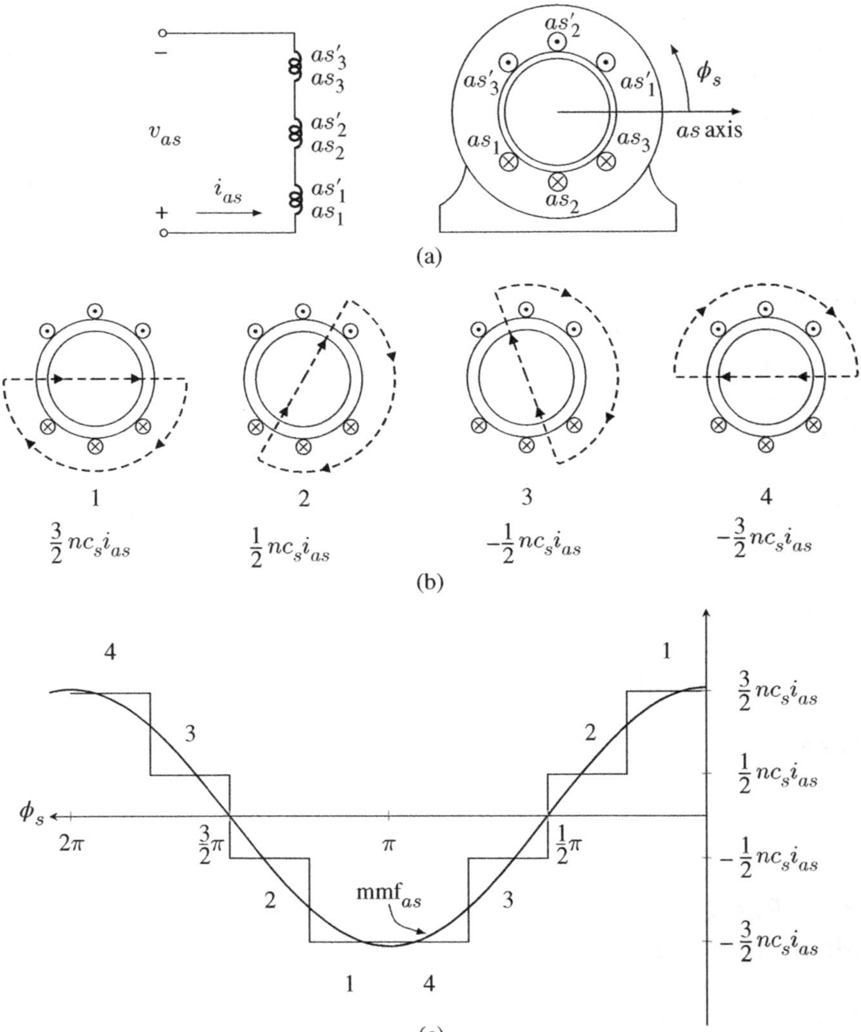

Figure 2C.1 Elementary two-pole single-phase stator winding over 90°. (a) Winding connections and distribution. (b) Ampere's law. (c) mmf_{as}.

2.5.1 Two-pole Two-phase Stator

In the case of a symmetrical two-pole two-phase stator, the *bs* winding is located $\frac{\pi}{2}$ radians from the *as* winding and it is identical in distribution and parameters to the *as* winding. Let us assume it is positioned as illustrated in Figure 2.9. Therein, a circle placed at the position of maximum turn density is now used to

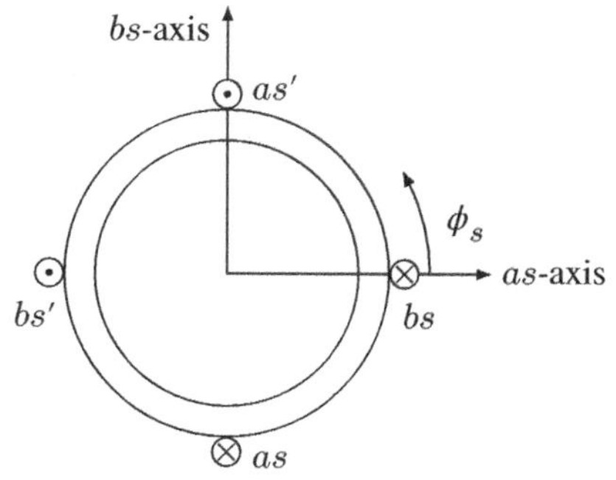

Figure 2.9 Elementary two-pole two-phase sinusoidally distributed stator windings.

represent a sinusoidally distributed winding. The air-gap mmf created by i_{bs} is approximated as

$$\mathrm{mmf}_{bs} = \frac{N_s}{2} i_{bs} \sin \phi_s \tag{2.38}$$

where N_s is the amplitude of the fundamental component of the winding distribution. Thus, the total air-gap mmf due to the two-pole stator windings would be

$$\begin{aligned}\mathrm{mmf}_s &= \mathrm{mmf}_{as} + \mathrm{mmf}_{bs} \\ &= \frac{N_s}{2}(i_{as} \cos \phi_s + i_{bs} \sin \phi_s)\end{aligned} \tag{2.39}$$

This relationship is perhaps the most important equation we'll run into in this text since it is key to reference frame theory and the operation and analysis of ac machines.

From Section 1.5, a two-phase steady-state balanced set of stator currents (*ab* sequence) may be expressed

$$I_{as} = \sqrt{2} I_s \cos[\omega_e t + \theta_{esi}(0)] \tag{2.40}$$

$$I_{bs} = \sqrt{2} I_s \sin[\omega_e t + \theta_{esi}(0)] \tag{2.41}$$

where I_s is the root-mean-square (rms) value of the phase current, ω_e is the electrical angular velocity in rad/s, and $\theta_{esi}(0)$ is the phase angle of the currents. For this balanced set $\tilde{I}_{bs} = -j\tilde{I}_{as}$.

Since Tesla's magnetic field rotates relative to the stator, perhaps, looking at it from a rotating axis will help us visualize what is going on. In Figure 2.10 we have added a third axis, the *q* axis, to Figure 2.9. We'll explain why we use "*q*" later. As shown in Figure 2.10, the concept of a *q* axis is no different from the *as* or *bs* axis, except that it can rotate at any angular velocity ω, or it can be stationary ($\omega = 0$). We are free to specify it. The angle from the *as* axis to the *q* axis is θ and for a constant ω

$$\theta = \omega t + \theta(0) \tag{2.42}$$

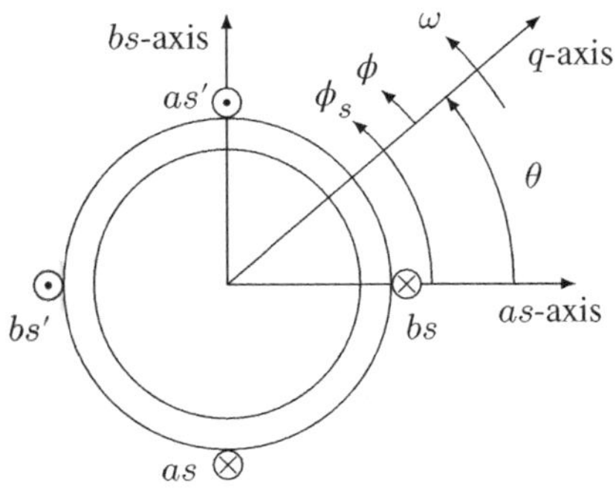

Figure 2.10 Elementary two-pole two-phase sinusoidally distributed stator windings with a third magnetic axis (q axis).

The displacement ϕ shown in Figure 2.10 is from the q axis just as ϕ_s is the displacement from the *as* axis. Let us relate a rotating position to an adjacent position on the stator. From Figure 2.10

$$\phi_s = \theta + \phi \tag{2.43}$$

If we substitute (2.43) into (2.39), we obtain

$$\text{mmf}_s = \frac{N_s}{2}[i_{as}\cos(\theta + \phi) + i_{bs}\sin(\theta + \phi)] \tag{2.44}$$

Substituting (2.40) and (2.41) into (2.44) for i_{as} and i_{bs}, respectively, and (2.42) for θ and after some work

$$\text{mmf}_s = \frac{N_s}{2}\sqrt{2}I_s\cos[(\omega_e - \omega)t + \theta_{esi}(0) - \theta(0) - \phi] \tag{2.45}$$

Recall that ω_e is the angular frequency of the electrical system and ω is the angular velocity of the q axis, $\theta(0)$ is the time zero position of the q axis which will be considered to be zero unless otherwise specified, and ϕ is the displacement from the q axis.

If we let the angular velocity of the q axis be zero ($\omega = 0$) and $\theta(0) = 0$, then $\phi = \phi_s$ and we would be viewing the rotating air-gap mmf, or, Tesla's rotating magnetic field as a stationary observer. In this case, (2.45) becomes

$$\text{mmf}_s^s = \frac{N_s}{2}\sqrt{2}I_s\cos[\omega_e t + \theta_{esi}(0) - \phi_s] \tag{2.46}$$

where we have added a superscript s to emphasize that we are observing Tesla's rotating magnetic field from a stationary frame of reference. Since both $\theta_{esi}(0)$ and ϕ_s are constants, mmf_s^s is a sinusoidal variation of frequency ω_e. If, for example, ϕ_s is zero we would be positioned (fixed) at the *as* axis ($\phi_s = 0$ in Figure 2.10) and the mmf_s^s would be pulsating at ω_e.

If we now let $\omega = \omega_e$, (2.45) becomes

$$\text{mmf}_s^e = \frac{N_s}{2}\sqrt{2}I_s\cos[\theta_{esi}(0) - \phi_e] \tag{2.47}$$

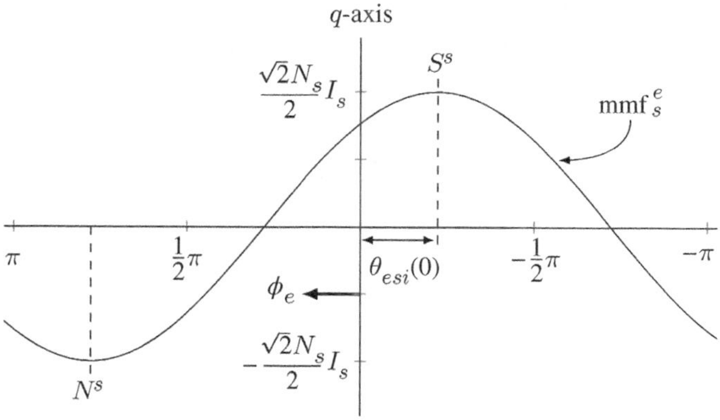

Figure 2.11 Tesla's rotating magnetic field (mmf_s^e) viewed from $-\pi$ to π by an observer rotating counterclockwise about the air gap at ω_e with $\theta(0) = 0$ and $\theta_{esi}(0)$ is negative.

where ϕ_e is the displacement from the q axis, which is rotating at ω_e. Since $\theta_{esi}(0)$ is constant for balanced steady-state conditions and let's say we have 360° vision, then (2.47) appears as a sinusoidal function of ϕ_e (the displacement from the q axis) with its maximum value occurring at $\phi_e = \theta_{esi}(0)$ which is the phase angle of the balanced steady-state stator currents. In other words, we are observing Tesla's rotating magnetic field as we run at ω_e in the counterclockwise direction around the air gap. Since ω_e is the angular frequency of I_{as} and I_{bs} and the q axis is also rotating at ω_e or in synchronism with the electrical system, it is referred to as the synchronously rotating frame of reference. This is depicted in Figure 2.11 where $\theta_{esi}(0)$ is negative for inductive circuits. A positive mmf_s^e is a south pole due to the stator currents (S^s) with the maximum intensity at $\phi_e = \theta_{esi}(0)$. A negative mmf_s^e represents a stator north pole (N^s).

We have just opened the door to a concept of viewing Tesla's rotating magnetic field from a frame of reference other than that of a stationary observer; in this case, a frame of reference rotating at ω_e in the counterclockwise direction. The importance of this cannot be overemphasized since we will find that this concept not only enables a clear visualization of the operation of electric machines, but it provides the basis for modern analysis of electric machines.

This is not the first time we have viewed a sinusoidal variation from a reference frame rotating at ω_e. Recall in Section 1.2, Example 1A, it was illustrated that a phasor rotated at ω_e in the counterclockwise direction and when we consider the phasor as a constant we are running counterclockwise at ω_e or we are stopping rotation at $\theta = 0$. We'll talk more of this in later sections; however, before proceeding there are a couple things that should be mentioned. It is important not to confuse the magnetic axes of the stator windings (as and bs axes) which are stationary, with the phasors representing the stator currents ($\tilde{I}_{as}$ and $\tilde{I}_{bs}$). However, it is interesting to note the relative position of the *as* axis and *bs* axis versus the relative position of $\tilde{I}_{as}$ and $\tilde{I}_{bs}$ in a phasor diagram for a constant, counterclockwise mmf_s to occur. The *bs* axis is displaced $\frac{1}{2}\pi$ ahead of the *as* axis, as illustrated in Figure 2.9; however, from (2.40) and (2.41) we see that $\tilde{I}_{bs}$ lags $\tilde{I}_{as}$ by $\frac{1}{2}\pi$. The direction of rotation of the air-gap mmf (mmf_s) may readily be determined by forgetting about $\theta_{esi}(0)$ and locating the position of $\frac{N_s}{2}\sqrt{2}I_s$ when time has progressed to where I_{bs} is maximum and I_{as} is zero $\left(\omega_e t = \frac{1}{2}\pi\right)$. That is, the maximum mmf_s has rotated counterclockwise from the *as* axis to the *bs* axis. Note that the negative of both (2.40) and (2.41) would also produce a counterclockwise mmf_s. In this case, $\tilde{I}_{as}$, would be at 180° and $\tilde{I}_{bs}$ at 90°. In other words, maximum I_{as} followed $\frac{\pi}{2}$ electrical degrees by maximum I_{bs} produces counterclockwise rotation of mmf_s in the device shown in Figure 2.9.

Example 2D ***Components of*** $\boldsymbol{mmf_s^e}$

Express mmf_{as}^e and mmf_{bs}^e with $\theta(0) = 0$. Add the resulting relationships to obtain mmf_s^e which is (2.47). The currents may be expressed

$$I_{as} = \sqrt{2}I_s \cos[\omega_e t + \theta_{esi}(0)] \tag{2D.1}$$

$$I_{bs} = \sqrt{2}I_s \sin[\omega_e t + \theta_{esi}(0)] \tag{2D.2}$$

To view the mmf as an observer running at ω_e counterclockwise with the q axis (Figure 2.10)

$$\phi_s = \omega_e t + \phi_e \tag{2D.3}$$

where ϕ_e is the displacement from the q axis when it is rotating at ω_e with $\theta(0) = 0$. Thus, from (2.44)

$$\begin{aligned}\text{mmf}_{as}^e &= \frac{N_s}{2}\sqrt{2}I_s\{\cos[\omega_e t + \theta_{esi}(0)]\cos(\omega_e t + \phi_e)\}\\ &= \frac{N_s}{2}\sqrt{2}I_s\left(\frac{1}{2}\right)\{\cos[2\omega_e t + \theta_{esi}(0) + \phi_e] + \cos[\theta_{esi}(0) - \phi_e]\}\end{aligned} \tag{2D.4}$$

From (2.38) and (2D.2), we can express mmf_{bs}^e as

$$\text{mmf}_{bs}^e = \frac{N_s}{2}\sqrt{2}I_s\left(\frac{1}{2}\right)\{\cos[\theta_{esi}(0) - \phi_e] - \cos[2\omega_e t + \theta_{esi}(0) + \phi_e]\} \tag{2D.5}$$

Now, from (2.39)

$$\begin{aligned}\text{mmf}_s^e &= \text{mmf}_{as}^e + \text{mmf}_{bs}^e\\ &= \frac{N_s}{2}\sqrt{2}I_s\cos[\theta_{esi}(0) - \phi_e]\end{aligned} \tag{2D.6}$$

The double-frequency terms cancel and we obtain (2.47) which is (2D.6).

2.5.2 Three-phase Stator

The arrangement of the stator windings of a two-pole three-phase device is shown in Figure 2.12. The windings are connected in wye and they are identical, sinusoidally distributed with N_s equivalent turns and with their magnetic axes displaced by $\frac{2}{3}\pi$. The positive direction of the magnetic axes is selected so as to achieve counterclockwise rotation of the rotating air-gap mmf with balanced stator currents of the *abc* sequence. We shall see this in just a moment. The air-gap mmfs established by the stator phases may be expressed by inspection of Figure 2.12. In particular,

$$\text{mmf}_{as} = \frac{N_s}{2} i_{as} \cos \phi_s \tag{2.48}$$

$$\text{mmf}_{bs} = \frac{N_s}{2} i_{bs} \cos \left(\phi_s - \frac{2}{3}\pi \right) \tag{2.49}$$

$$\text{mmf}_{cs} = \frac{N_s}{2} i_{cs} \cos \left(\phi_s + \frac{2}{3}\pi \right) \tag{2.50}$$

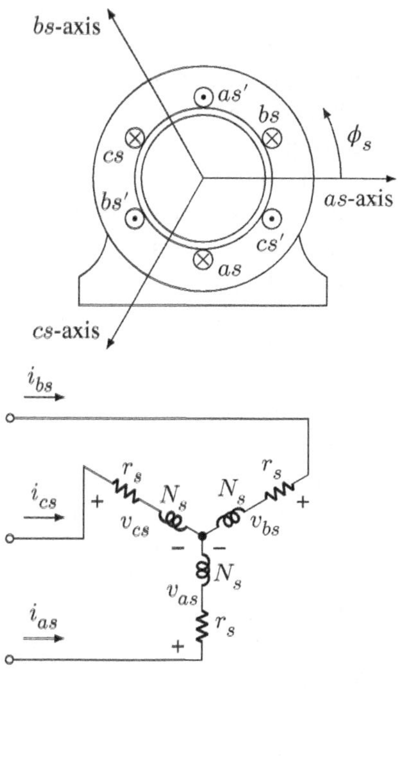

Figure 2.12 Elementary two-pole three-phase sinusoidally distributed stator windings.

As before, N_s is the number of turns of the equivalent sinusoidally distributed stator windings and ϕ_s is the angular displacement about the stator measured from the *as* axis. The total air-gap mmf_s is

$$\text{mmf}_s = \frac{N_s}{2}\left[i_{as}\cos\phi_s + i_{bs}\cos\left(\phi_s - \frac{2}{3}\pi\right) + i_{cs}\cos\left(\phi_s + \frac{2}{3}\pi\right)\right] \tag{2.51}$$

For balanced steady-state conditions, the stator currents for an *abc* sequence may be expressed as

$$I_{as} = \sqrt{2}I_s\cos[\omega_e t + \theta_{esi}(0)] \tag{2.52}$$

$$I_{bs} = \sqrt{2}I_s\cos\left[\omega_e t - \frac{2}{3}\pi + \theta_{esi}(0)\right] \tag{2.53}$$

$$I_{cs} = \sqrt{2}I_s\cos\left[\omega_e t + \frac{2}{3}\pi + \theta_{esi}(0)\right] \tag{2.54}$$

Substituting (2.52)–(2.54) into (2.51) and using (2.43) for ϕ_s and then after some trigonometric manipulations, we can obtain an expression for the rotating air-gap mmf (Tesla's rotating field) established by balanced steady-state stator currents

$$\text{mmf}_s = \frac{N_s}{2}\sqrt{2}I_s\frac{3}{2}\cos[(\omega_e - \omega)t + \theta_{esi}(0) - \theta(0) - \phi] \tag{2.55}$$

If mmf_s for the three-phase device given by (2.55) is compared with mmf_s for a two-phase device given by (2.45), we see that they are identical except that the amplitude for the three-phase device is $\frac{3}{2}$ times that of a two-phase device. Actually, it can be shown that the amplitude for multiphase devices changes from that of a two-phase device in proportion to the number of phases divided by two. It has been pointed out that we will first analyze the two-phase machine since the trigonometry is less involved and the extension to the three-phase machine, once we have considered its two-phase counterpart, is straightforward. It follows that we need not repeat the work we did leading up to Figure 2.11. The only difference would be that (2.45) would be multiplied by $\frac{3}{2}$.

It is important to note that, with the assumed positive directions of the magnetic axes, a counterclockwise rotating air-gap mmf is obtained with a three-phase set of balanced stator currents of the *abc* sequence. As in the two-phase case, it is also important to note the relative positions of the magnetic axes versus the relative positions of the phasors representing the currents to establish a constant amplitude counterclockwise rotating air-gap mmf. From (2.48)–(2.50) or Figure 2.12 we see that the *bs* axis is stationary at 120°, whereas the *cs* axis is stationary at −120°. From (2.52)–(2.54), $\tilde{I}_{as}$, $\tilde{I}_{bs}$, and $\tilde{I}_{cs}$ are 120° out of phase; however, for the constant-amplitude air-gap mmf to rotate in the counterclockwise direction, $\tilde{I}_{bs}$ lags $\tilde{I}_{as}$ by 120° and $\tilde{I}_{cs}$ lags $\tilde{I}_{as}$ by 240°. In other words, maximum positive I_{as} is followed by maximum I_{bs} and then maximum I_{cs} for an *abc* sequence. With the magnetic axis positioned as shown in Figure 2.12, this sequence of maximum currents produces a counterclockwise rotation of the air-gap mmf.

SP2.5.1 We have expressed mmf_s^s and mmf_s^e. Express mmf_s^r; a reference frame rotating counterclockwise at the angular velocity of the rotor, ω_r. [Equation (2.45) or (2.55) with $\omega = \omega_r$ and ϕ replaced by ϕ_r]

SP2.5.2 Express mmf_s^s for a three-phase stator. [Equation (2.46) multiplied by $\frac{3}{2}$]

2.6 Change of Variables

R. H. Park came up with a change of variables in 1929 that eliminated the position varying inductances of a synchronous machine. His work revolutionized machine analysis [2]. Only recently was the connection between Tesla and Park discovered [3]. As it turns out, the change of variables – *reference frame theory*, as it is often called – is nothing more than establishing the variables associated with the fictitious circuits that produce Tesla's rotating magnetic field as viewed from the particular reference frame of interest.

2.6.1 Two-phase Transformation

The transformation or change of variables can be readily established from (2.39), (2.43), and (2.44). These equations are repeated here to start our derivation.

$$\text{mmf}_s = \frac{N_s}{2}(i_{as}\cos\phi_s + i_{bs}\sin\phi_s) \tag{2.56}$$

$$\phi_s = \theta + \phi \tag{2.57}$$

Substituting (2.57) into (2.56) yields

$$\text{mmf}_s = \frac{N_s}{2}[i_{as}\cos(\theta+\phi) + i_{bs}\sin(\theta+\phi)] \tag{2.58}$$

Equation (2.58) may be written, using basic trigonometry, as

$$\text{mmf}_s = \frac{N_s}{2}\cos\phi(i_{as}\cos\theta + i_{bs}\sin\theta) + \frac{N_s}{2}\sin\phi(-i_{as}\sin\theta + i_{bs}\cos\theta) \tag{2.59}$$

As shown in Figure 2.13, $N_s \cos\phi$ and $N_s \sin\phi$ result from two orthogonal sinusoidally distributed windings.

The current flowing in the winding "hovered" around the q axis ($\phi = 0$) is i_{qs} and it is expressed as

$$i_{qs} = i_{as}\cos\theta + i_{bs}\sin\theta \tag{2.60}$$

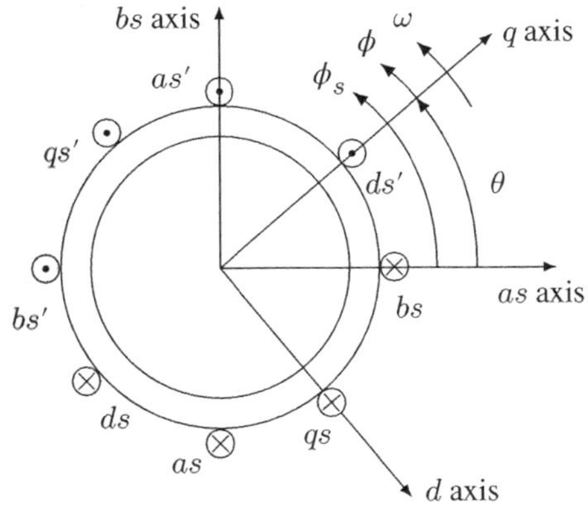

Figure 2.13 Elementary two-pole two-phase stator with q and d axes.

The second winding can be at $\phi = \pm\frac{\pi}{2}$. We'll let $\phi = -\frac{\pi}{2}$, then the current which we'll call i_{ds} is expressed

$$i_{ds} = i_{as} \sin\theta - i_{bs} \cos\theta \tag{2.61}$$

we can now write

$$\text{mmf}_s = \frac{N_s}{2}(i_{qs} \cos\phi + i_{ds} \sin\phi) \tag{2.62}$$

The change of variables expressed in matrix form is

$$\begin{bmatrix} i_{qs} \\ i_{ds} \end{bmatrix} = \begin{bmatrix} \cos\theta & \sin\theta \\ \sin\theta & -\cos\theta \end{bmatrix} \begin{bmatrix} i_{as} \\ i_{bs} \end{bmatrix} \tag{2.63}$$

This same transformation also applies to voltage and flux linkages, thus

$$\begin{bmatrix} f_{qs} \\ f_{ds} \end{bmatrix} = \begin{bmatrix} \cos\theta & \sin\theta \\ \sin\theta & -\cos\theta \end{bmatrix} \begin{bmatrix} f_{as} \\ f_{bs} \end{bmatrix} \tag{2.64}$$

Where f can be voltage, current, or flux linkage. We can also write (2.63) as

$$\mathbf{f}_{qds} = \mathbf{K}_s \mathbf{f}_{abs} \tag{2.65}$$

Now,

$$\frac{d\theta}{dt} = \omega(t) \tag{2.66}$$

where ω is the unspecified speed of the reference frame; we are free to assign it any value and $\mathbf{K}_s$ is considered as the transformation to the arbitrary reference frame. In this text we will only concern ourselves with three specific reference frames; the stationary, $\omega = 0$, the rotor, $\omega = \omega_r$, and the synchronously rotating, $\omega = \omega_e$. We'll talk more of this as we go along.

2.6.2 Three-phase Transformation

For the three-phase transformation, we'll start with (2.51) which is

$$\text{mmf}_s = \frac{N_s}{2}\left[i_{as}\cos\phi_s + i_{bs}\cos\left(\phi_s - \frac{2}{3}\pi\right) + i_{cs}\cos\left(\phi_s + \frac{2}{3}\pi\right)\right] \tag{2.67}$$

Substituting (2.57) for ϕ_s yields

$$\begin{aligned}\text{mmf}_s = &\frac{N_s}{2}\cos\phi\left[i_{as}\cos\phi + i_{bs}\cos\left(\phi - \frac{2}{3}\pi\right) + i_{cs}\cos\left(\phi + \frac{2}{3}\pi\right)\right]\\ &- \frac{N_s}{2}\sin\phi\left[i_{as}\sin\theta + i_{bs}\sin\left(\theta - \frac{2}{3}\pi\right) + i_{cs}\sin\left(\theta + \frac{2}{3}\pi\right)\right]\end{aligned} \tag{2.68}$$

Here i_{qs} is the [ ] multiplying $\cos\phi$ and $-i_{ds}$ is the [ ] multiplying $\sin\phi$ and (2.62) also applies. In addition to the q and d variables a third variable, the zero variable, is introduced as

$$f_{0s} = \frac{1}{3}(f_{as} + f_{bs} + f_{cs}) \tag{2.69}$$

The transformation to the arbitrary reference frame becomes

$$\mathbf{f}_{qd0s} = \mathbf{K}_s\mathbf{f}_{abcs} \tag{2.70}$$

where in the case of the three-phase machine

$$(\mathbf{f}_{qd0s})^T = [f_{qs}\ f_{ds}\ f_{0s}] \tag{2.71}$$

$$(\mathbf{f}_{abcs})^T = [f_{as}\ \ f_{bs}\ \ f_{cs}] \tag{2.72}$$

$$\mathbf{K}_s = \frac{2}{3}\begin{bmatrix}\cos\theta & \cos\left(\theta - \frac{2}{3}\pi\right) & \cos\left(\theta + \frac{2}{3}\pi\right)\\ \sin\theta & \sin\left(\theta - \frac{2}{3}\pi\right) & \sin\left(\theta + \frac{2}{3}\pi\right)\\ \frac{1}{2} & \frac{1}{2} & \frac{1}{2}\end{bmatrix} \tag{2.73}$$

$$(\mathbf{K}_s)^{-1} = \begin{bmatrix}\cos\theta & \sin\theta & 1\\ \cos\left(\theta - \frac{2}{3}\pi\right) & \sin\left(\theta - \frac{2}{3}\pi\right) & 1\\ \cos\left(\theta + \frac{2}{3}\pi\right) & \sin\left(\theta + \frac{2}{3}\pi\right) & 1\end{bmatrix} \tag{2.74}$$

where

$$\frac{d\theta}{dt} = \omega(t) \tag{2.75}$$

Note that the f_{0s} variable does not contain θ_r and is zero for balanced conditions. The $\frac{2}{3}$ factor in (2.72) was introduced by Park. It is not suggested by Tesla. Due to this $\frac{2}{3}$ factor in Park's transformation, which makes the magnitude of the variables equal to those of the two-phase system, we must multiply the power and torque by $\frac{3}{2}$ to correct this reduction in the power calculated using Park's variables. We'll take care of all this when we consider the individual machines.

Example 2E ***Applied voltages in different reference frames***
We will assume that

$$v_{as} = \sqrt{2}V_s \cos\theta_{esv} \tag{2E.1}$$

$$v_{bs} = \sqrt{2}V_s \sin\theta_{esv} \tag{2E.2}$$

$$\theta_{esv} = \omega_e t + \theta_{esv}(0) \tag{2E.3}$$

Throughout this text the applied voltages will be of the form given by (2E.1)–(2E.3). Although V_s and θ_{esv} are generally constant, it is desirable to express the applied voltages, v_{qs} and v_{ds} in the arbitrary, stationary, and synchronously rotating reference frames. For this purpose, let

$$\theta = \omega t + \theta(0) \tag{2E.4}$$

Substituting (2E.1)–(2E.4) into (2.63) and after some work we have the expression for the applied voltages in the arbitrary reference frame

$$v_{qs} = \sqrt{2}V_s \cos[(\omega_e - \omega)t + \theta_{esv}(0) - \theta(0)] \tag{2E.5}$$

$$v_{ds} = -\sqrt{2}V_s \sin[(\omega_e - \omega)t + \theta_{esv}(0) - \theta(\theta)] \tag{2E.6}$$

where $\theta_{esv}(0)$ is the phase angle of v_{as} and v_{bs} and $\theta(0)$ is zero unless otherwise specified.

For the stationary reference frame $\omega = 0$ and

$$v_{qs}^s = v_{as} \tag{2E.7}$$

$$v_{ds}^s = -v_{bs} \tag{2E.8}$$

where the superscript s denotes variables in the stationary reference frame. In the synchronously rotating reference frame, $\omega = \omega_e$

$$v_{qs}^e = \sqrt{2}V_s \cos\theta_{esv}(0) \tag{2E.9}$$

$$v_{ds}^e = -\sqrt{2}V_s \sin\theta_{esv}(0) \tag{2E.10}$$

where the raised e denotes variables in the synchronously rotating reference frame. It is important to note that, in the steady state, v_{qs}^e and v_{ds}^e are constant dc, applied voltages. It is a good bet that the currents i_{qs}^e and i_{ds}^e will also be dc in the steady state.

SP2.6.1 Express v_{qs} and v_{ds} in the arbitrary reference frame for a balanced three-phase system. The trigonometric identities in Appendix A will help. [Equations (2E.5) and (2E.6)]

SP2.6.2 Assume v_{ds} and i_{ds} are both zero. Show that we must multiply power calculated using Parks transformation by $\frac{3}{2}$. $[P_{3\phi} = \frac{3}{2}P_{2\phi} = \left(\frac{3}{2}\right) 2V_sI_s = 3V_sI_s]$

SP2.6.3 Show that for balanced conditions f_{0s} is zero. $[f_{as} + f_{bs} + f_{cs} = 0]$

SP2.6.4 Use (2E.5) and (2E.6) to express v^r_{qs} and v^r_{ds}. [Let $\omega = \omega_r$]

2.7 Stator Voltage Equations in Arbitrary Reference Frame

We will now transform the *as* and *bs* variables to the arbitrary reference frame. To do this we'll consider Figure 2.13, where

$$v_{as} = r_s i_{as} + \frac{d\lambda_{as}}{dt} \tag{2.76}$$

$$v_{bs} = r_s i_{bs} + \frac{d\lambda_{bs}}{dt} \tag{2.77}$$

which may be written

$$\mathbf{v}_{abs} = \mathbf{r}_s \mathbf{i}_{abs} + p\boldsymbol{\lambda}_{abs} \tag{2.78}$$

From (2.64)

$$\mathbf{f}_{abs} = (\mathbf{K}_s)^{-1}\mathbf{f}_{qds} \tag{2.79}$$

Substituting into (2.78) yields

$$(\mathbf{K}_s)^{-1}\mathbf{v}_{qds} = \mathbf{r}_s(\mathbf{K}_s)^{-1}\mathbf{i}_{qds} + p[(\mathbf{K}_s)^{-1}\boldsymbol{\lambda}_{qds}] \tag{2.80}$$

Multiplying each side of (2.80) by $\mathbf{K}_s$, gives

$$\mathbf{v}_{qds} = \mathbf{K}_s\mathbf{r}_s(\mathbf{K}_s)^{-1}\mathbf{i}_{qds} + \mathbf{K}_s p[(\mathbf{K}_s)^{-1}]\boldsymbol{\lambda}_{qds} + \mathbf{K}_s(\mathbf{K}_s)^{-1}p\boldsymbol{\lambda}_{qds} \tag{2.81}$$

Now, term by term

$$\begin{aligned}\mathbf{K}_s\mathbf{r}_s(\mathbf{K}_s)^{-1}\mathbf{i}_{qds} &= \mathbf{K}_s r_s \mathbf{I}(\mathbf{K}_s)^{-1}\mathbf{i}_{qds}\\ &= r_s\mathbf{K}_s(\mathbf{K}_s)^{-1}\mathbf{i}_{qds} = \mathbf{r}_s\mathbf{i}_{qds}\end{aligned} \tag{2.82}$$

$$\begin{aligned}\mathbf{K}_s p(\mathbf{K}_s)^{-1}\boldsymbol{\lambda}_{qds} &= \mathbf{K}_s\omega\begin{bmatrix}-\sin\theta & \cos\theta\\ \cos\theta & \sin\theta\end{bmatrix}\begin{bmatrix}\lambda_{qs}\\ \lambda_{ds}\end{bmatrix}\\ &= \omega\begin{bmatrix}0 & 1\\ -1 & 0\end{bmatrix}\begin{bmatrix}\lambda_{qs}\\ \lambda_{ds}\end{bmatrix}\end{aligned}$$

$$= \omega \begin{bmatrix} \lambda_{ds} \\ -\lambda_{qs} \end{bmatrix} = \omega \boldsymbol{\lambda}_{dqs} \tag{2.83}$$

$$\mathbf{K}_s(\mathbf{K}_s)^{-1} p\boldsymbol{\lambda}_{qds} = p\boldsymbol{\lambda}_{qds} \tag{2.84}$$

The voltage equation in the arbitrary reference frame becomes

$$\mathbf{v}_{qds} = \mathbf{r}_s \mathbf{i}_{qds} + \omega \boldsymbol{\lambda}_{dqs} + p\boldsymbol{\lambda}_{qds} \tag{2.85}$$

where $\omega \boldsymbol{\lambda}_{dqs}$ is (2.83). In expanded form

$$v_{qs} = r_s i_{qs} + \omega \lambda_{ds} + p\lambda_{qs} \tag{2.86}$$

$$v_{ds} = r_s i_{ds} - \omega \lambda_{qs} + p\lambda_{ds} \tag{2.87}$$

The voltage equations in the stationary reference frame become, ($\omega = 0$)

$$v_{qs}^s = r_s i_{qs}^s + p\lambda_{qs}^s \tag{2.88}$$

$$v_{ds}^s = r_s i_{ds}^s + p\lambda_{ds}^s \tag{2.89}$$

where the raised s denotes variables in the stationary reference frame. In the rotor reference frame, ($\omega = \omega_r$)

$$v_{qs}^r = r_s i_{qs}^r + \omega_r \lambda_{ds}^r + p\lambda_{qs}^r \tag{2.90}$$

$$v_{ds}^r = r_s i_{ds}^r - \omega_r \lambda_{qs}^r + p\lambda_{ds}^r \tag{2.91}$$

where the raised r denotes variables in the rotor reference frame. In the synchronous rotating reference frame, ($\omega = \omega_e$)

$$v_{qs}^e = r_s i_{qs}^e + \omega_e \lambda_{ds}^e + p\lambda_{qs}^e \tag{2.92}$$

$$v_{ds}^e = r_s i_{ds}^e - \omega_e \lambda_{qs}^e + p\lambda_{ds}^e \tag{2.93}$$

where the raised e denotes variables in the synchronously rotating reference frame. These equations provide the rotating magnetic field as viewed from the reference frames. The applied voltages in each reference frame are given for balanced conditions in Example 2E.

We have not defined λ_{qs} and λ_{ds}; therefore, the equations derived in this section are valid for linear and nonlinear magnetic systems. We'll set forth the flux linkage equations when we consider each machine whereupon we will express flux linkages in terms of inductances making the equations valid only for linear systems.

Example 2F ***Voltage equations for $\boldsymbol{\phi_s = \frac{\pi}{2}}$ for $\mathbf{i}_{ds}$***

The transformation for i_{ds} at $\phi_s = \frac{\pi}{2}$ is (see SP2.7.3)

$$\begin{bmatrix} f_{qs} \\ f_{ds} \end{bmatrix} = \begin{bmatrix} \cos\theta & \sin\theta \\ -\sin\theta & \cos\theta \end{bmatrix} \begin{bmatrix} f_{as} \\ f_{bs} \end{bmatrix} \tag{2F.1}$$

Now, (2.82) and (2.84) do not change, that is

$$\mathbf{K}_s\mathbf{r}_s(\mathbf{K}_s)^{-1}\mathbf{i}_{qds} = \mathbf{r}_s\mathbf{i}_{qds} \tag{2F.2}$$

$$\mathbf{K}_s(\mathbf{K}_s)^{-1}p\boldsymbol{\lambda}_{qds} = p\boldsymbol{\lambda}_{qds} \tag{2F.3}$$

The speed voltage changes

$$\begin{aligned}
\mathbf{K}_s p(\mathbf{K}_s)^{-1}\boldsymbol{\lambda}_{qds} &= \mathbf{K}_s\omega \begin{bmatrix} -\sin\theta & -\cos\theta \\ \cos\theta & -\sin\theta \end{bmatrix} \begin{bmatrix} \lambda_{qs} \\ \lambda_{ds} \end{bmatrix} \\
&= \omega \begin{bmatrix} \cos\theta & \sin\theta \\ -\sin\theta & \cos\theta \end{bmatrix} \begin{bmatrix} -\sin\theta & -\cos\theta \\ \cos\theta & -\sin\theta \end{bmatrix} \begin{bmatrix} \lambda_{qs} \\ \lambda_{ds} \end{bmatrix} \\
&= \omega \begin{bmatrix} 0 & -1 \\ 1 & 0 \end{bmatrix} \begin{bmatrix} \lambda_{qs} \\ \lambda_{ds} \end{bmatrix} \\
&= \omega \begin{bmatrix} -\lambda_{ds} \\ \lambda_{qs} \end{bmatrix}
\end{aligned} \tag{2F.4}$$

The voltage equations in the arbitrary reference frame become

$$v_{qs} = r_s i_{qs} - \omega\lambda_{ds} + p\lambda_{qs} \tag{2F.5}$$

$$v_{ds} = r_s i_{ds} + \omega\lambda_{qs} + p\lambda_{ds} \tag{2F.6}$$

The sign associated with the speed voltages is reversed.

SP2.7.1 Let $V_{as} = \sqrt{2}V_s\sin[\omega_e t + \theta_{esv}(0)], V_{bs} = -\sqrt{2}V_s\cos[\omega_e t + \theta_{esv}(0)]$. Express V_{qs}^e and V_{ds}^e. [$V_{qs}^e = \sqrt{2}V_s\sin\theta_{esv}(0)$, $V_{ds}^e = \sqrt{2}V_s\cos\theta_{esv}(0)$]

SP2.7.2 Verify (2.83).

SP2.7.3 Determine $(\mathbf{K}_s)^{-1}$ for $\mathbf{K}_s$ with i_{ds} from (2F.1). [$(\mathbf{K}_s)^{-1} = \begin{bmatrix} \cos\theta & -\sin\theta \\ \sin\theta & \cos\theta \end{bmatrix}$; $\mathbf{K}_s \neq (\mathbf{K}_s)^{-1}$]

2.8 Instantaneous and Steady-state Phasors

The synchronously rotating reference frame can be thought of as a rotating complex plane as shown in Figure 2.14 whereupon we can think of f_{qs}^e and f_{ds}^e as $\tilde{f}_{as}$, an instantaneous phasor of phase *as* variables. This would include the steady-state and transient response for balanced conditions. From Figure 2.14, we can write

$$\tilde{f}_{as} = f_{qs}^e - jf_{ds}^e \tag{2.94}$$

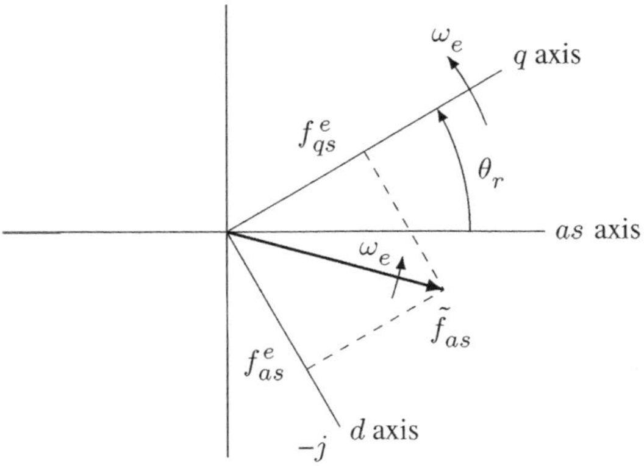

Figure 2.14 The q and d complex plane.

Substituting (2.92) and (2.93) into (2.94) where upon

$$v_{qs}^e - jv_{ds}^e = r_s i_{qs}^e + \omega_e \lambda_{ds}^e + p\lambda_{qs}^e - j(r_s i_{ds}^e - \omega_e \lambda_{qs}^e + p\lambda_{ds}^e) \tag{2.95}$$

In terms of instantaneous phasors

$$\begin{aligned}\tilde{v}_{as} &= r_s(i_{qs}^e - ji_{ds}^e) + \omega_e(\lambda_{ds}^e + j\lambda_{qs}^e) + p(\lambda_{qs}^e - j\lambda_{ds}^e) \\ &= r_s\tilde{i}_{as} + j\omega_e\tilde{\lambda}_{as} + p\tilde{\lambda}_{as}\end{aligned} \tag{2.96}$$

Once the transient subsides, the i_{qs}^e and i_{ds}^e become constant and $p\tilde{\lambda}_{as}$ becomes zero and (2.96) becomes

$$\tilde{v}_{as} = r_s\tilde{i}_{as} + j\omega_e\tilde{\lambda}_{as} \tag{2.97}$$

Now, (2.97) is instantaneous values and must be divided by $\sqrt{2}$ to express in terms of steady-state phasors. Also, for a magnetically linear system, for the *rL* circuit being considered

$$\tilde{\lambda}_{as} = L\tilde{i}_{as} \tag{2.98}$$

thus (2.97) becomes

$$\tilde{V}_{as} = (r + j\omega_e L)\tilde{I}_{as} \tag{2.99}$$

and

$$\sqrt{2}\tilde{F}_{as} = F_{qs}^e - jF_{ds}^e \tag{2.100}$$

where the uppercase letters denote steady state. Equation (2.100) gives the relation between F_{qs}^e and F_{ds}^e and $\tilde{F}_{as}$ since both the synchronous reference frame and the phasor are rotated at ω_e. We will make extensive use of the instantaneous phasor when dealing with the dynamic response of machine performance.

Example 2G ***The instantaneous and steady-state phasors***
Let $r_s = 3.4\ \Omega$, $L_{ls} = 1.1$ mH, $L_{ms} = 11$ mH. Assume $\omega_e = 377$ rad/s and

$$\lambda_{qs}^e = L_{ss} i_{qs}^e \tag{2G.1}$$

$$\lambda_{ds}^e = L_{ss} i_{ds}^e \tag{2G.2}$$

Express $\tilde{v}_{as}$ and $\tilde{V}_{as}$. From (2.96)

$$\begin{aligned} \tilde{v}_{as} &= 3.4\tilde{i}_{as} + j(377)(11 + 1.1) \times 10^{-3}\tilde{i}_{as} + p(11 + 1.1) \times 10^{-3}\tilde{i}_{as} \\ &= 3.4\tilde{i}_{as} + j4.6\tilde{i}_{as} + 0.012p\tilde{i}_{as} \end{aligned} \tag{2G.3}$$

When the system reaches steady state, $p\tilde{i}_{as}$ becomes zero and (2G.3) becomes

$$\tilde{v}_{as} = (3.4 + j4.6)\tilde{i}_{as} \tag{2G.4}$$

In terms of a steady-state phasor

$$\sqrt{2}\tilde{V}_{as} = (3.4 + j4.6)\sqrt{2}\tilde{I}_{as} \tag{2G.5}$$

or

$$\tilde{V}_{as} = (3.4 + j4.6)\tilde{I}_{as} \tag{2G.6}$$

SP2.8.1 Express $\tilde{f}_{bs}$ in terms of f_{qs}^e and f_{ds}^e. $[\tilde{f}_{bs} = -j(f_{qs}^e - jf_{ds}^e)]$

SP2.8.2 Add $\tilde{f}_{bs}$ to Figure 2.14. $[-j\tilde{f}_{as}]$

2.9 *P*-pole Machines

Thus far, we have considered only two-pole electromechanical motion devices; however, electromechanical devices may have any even number of poles (2, 4, 6, 8, ...) up to more than 50 in the case of large hydroturbine generators. We may at first be a little hesitant to tackle the analysis of machines with more than two poles, since it appears that this may complicate matters before we even get started. Fortunately, this is not the case. We will find that with a simple change of variables we can analyze all machines as if they were two-pole devices. We need only to modify the expression for evaluating torque and show that, physically, the actual rotor speed of a machine with more than two poles will be a multiple less than that which we determine from our two-pole analysis.

The rotating air-gap mmf of a machine with more than two poles can be determined by considering the four-pole device shown in Figure 2.15. In this figure, each phase winding consists of two series-connected windings, each of which is assumed to be sinusoidally distributed. For example, $as1'$ represents a group

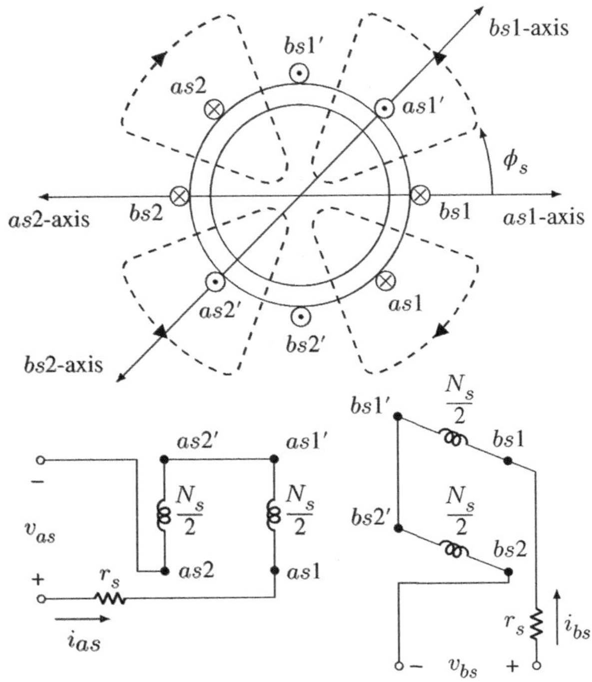

Figure 2.15 Stator winding arrangement of a four-pole, two-phase symmetrical electromechanical device. Flux streamlines shown for I_{as} maximum positive and I_{bs} is zero.

of conductors sinusoidally distributed over $0 < \phi_s < \frac{1}{2}\pi$, and so on. The phase windings consist of N_s turns, with $N_s/\frac{P}{2}$ turns in each of the two series-connected windings. There may be some confusion in regard to notation. In Figure 2.15, the notation *as*1, *bs*1, … is used to denote sinusoidally distributed windings; however, a subscripted number denotes a coil.

Note that each of the two series windings per phase spans $\frac{1}{2}\pi$ radians and each phase winding establishes two magnetic systems. As shown in Figure 2.15, at the instant when $I_{bs} = 0$ and I_{as} is a positive maximum, the $as1 - as1'$ part of the *as* winding produces positive flux in the *as*1-axis direction, whereas the $as2 - as2'$ part of the *as* winding produces positive flux in the *as*2-axis direction. South poles occur from $-\frac{1}{4}\pi < \phi_s < \frac{1}{4}\pi$ and $\frac{3}{4}\pi < \phi_s < \frac{5}{4}\pi$. Now, as shown in Figure 2.15 by the streamlines, half of the flux that enters the stator steel from $-\frac{1}{4}\pi < \phi_s < \frac{1}{4}\pi$ reenters the air gap between $\frac{5}{4}\pi < \phi_s < \frac{7}{4}\pi$ and the other half between $\frac{1}{4}\pi < \phi_s < \frac{3}{4}\pi$. The flux that enters the stator from $\frac{3}{4}\pi < \phi_s < \frac{5}{4}\pi$ divides similarly. Hence, for $I_{as} = \sqrt{2}I_s$ and $I_{bs} = 0$, two north poles occur; one from $\frac{1}{4}\pi < \phi_s < \frac{3}{4}\pi$ and from $\frac{5}{4}\pi < \phi_s < \frac{7}{4}\pi$.

The air-gap mmf established by each phase is a sinusoidal function of $2\phi_s$ for a four-pole machine, or, in general, $(P/2)\phi_s$, where P is the number of poles.

In particular,

$$\text{mmf}_{as} = \frac{2}{P}\frac{N_s}{2} i_{as} \cos \frac{P}{2}\phi_s \tag{2.101}$$

$$\text{mmf}_{bs} = \frac{2}{P}\frac{N_s}{2} i_{bs} \sin \frac{P}{2}\phi_s \tag{2.102}$$

where N_s is the total equivalent turns per stator phase. For balanced steady-state operation, the stator currents may be expressed as

$$I_{as} = \sqrt{2}I_s \cos[\omega_e t + \theta_{esi}(0)] \tag{2.103}$$

$$I_{bs} = \sqrt{2}I_s \sin[\omega_e t + \theta_{esi}(0)] \tag{2.104}$$

Equations (2.103) and (2.104) are the same expressions for the stator currents as given for the two-pole case. Also, (2.46) would become

$$\text{mmf}_s^s = \left(\frac{2}{P}\right)\frac{N_s}{2}\sqrt{2}I_s \cos\left[\omega_e t + \theta_{esi}(0) - \frac{P}{2}\phi_s\right] \tag{2.105}$$

With the stator arranged as in Figure 2.15, balanced steady-state stator currents of frequency ω_e produce a four-pole (P-pole) magnetic system that rotates about the air gap at $(2/4)\omega_e$ or $(2/P)\omega_e$ relative to a stationary observer. How do we know that? Well for (2.105) to be a constant, the argument of the cosine must be a constant and this occurs when $\frac{P}{2}\phi_s$ is equal to $\omega_e t$. In other words, when $\phi_s = \frac{2}{P}\omega_e t$. Therefore, the speed of the rotating mmf (rotating poles) is now $(2/P)\omega_e$ which is the synchronous speed of the machine; however, the stator variables are unaware of this. To the electric system, ω_e is synchronous speed; however, the magnetic system of the rotor must "lock in" with the magnetic system established by the stator.

A four-pole stator winding and a four-pole rotor winding are shown in Figure 2.16. The mutual inductance between the *as*- and *ar*-windings may be expressed

$$L_{asar} = L_{sr} \cos 2\theta_{rm} \tag{2.106}$$

where θ_{rm} is the actual rotor displacement. In the two-pole case, Figure 2.1, we used θ_r. We're going to give a new meaning to θ_r in just a minute. Let us first generalize the argument of the cosine in (2.106). Thus,

$$L_{asar} = L_{sr} \cos \frac{P}{2}\theta_{rm} \tag{2.107}$$

If we let

$$\frac{P}{2}\theta_{rm} = \theta_r \tag{2.108}$$

then (2.107) becomes what we had for the two-pole case. Now, θ_{rm} is the actual rotor displacement and θ_r is what is referred to as the electrical angular position

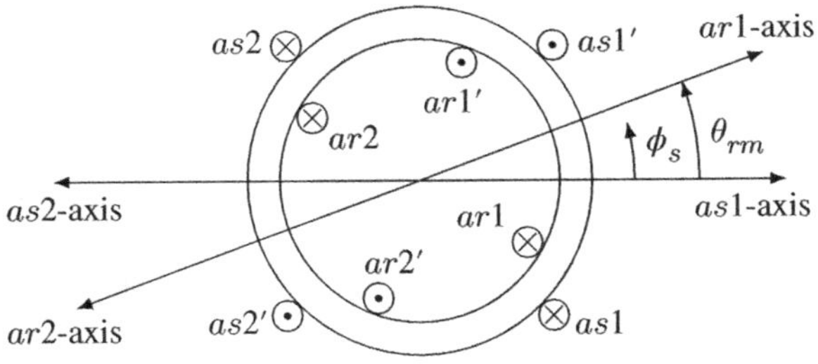

Figure 2.16 Mutual coupling between four-pole stator and rotor windings.

of the rotor. What in the world does that mean? Well, let us be stator variables; we don't know or care how many pole pairs the machine has, so we'll assume that both the stator and rotor have two poles. Thus, the electrical system is rotating at ω_e and it thinks the rotor speed is ω_r since we assumed it to be a two-pole stator and rotor. Therefore, ω_r is now the rotor speed "referred" to the angular velocity of the electrical system and (2.108) applies regardless of how many even-number poles are on the stator and rotor.

Although we see that if we substitute (2.108) into (2.107), the result is the mutual inductance for a two-pole system as given by (2.9); however, what about the mutual inductance between stator windings as shown in Figure 2.15? In the case of the two-pole machine, $L_{asbs} = 0$, since the magnetic axes are orthogonal. Note, however, that in the four-pole machine in Figure 2.15 the *bs* winding is displaced 45° counterclockwise from the *as* winding. If we assume, for a minute, that in Figure 2.15, we rotated the *bs*-winding clockwise 45°, it would be atop the *as*-winding and L_{asbs} would be similar to (2.107) except L_{asar} would be replaced by L_{asbs}, L_{sr} would be replaced by L_{ms} and θ_{rm} by the angle between *as* and *bs* windings. If now we rotate the *bs*-winding counterclockwise by an angle of 45°, then L_{asbs} would be zero. From here on out, we will treat all electric machines as two-pole devices, knowing fully well that (2.108) is the relationship between θ_r and θ_{rm} and

$$\frac{P}{2}\omega_{rm} = \omega_r \tag{2.109}$$

where ω_r is referred to as the electrical angular velocity of the rotor.

Now, since we are going to treat all machines as two-pole devices and use θ_r in all the voltage equations, we have to do one last thing to make things work out. Back in Section 2.3, we expressed the change in energy entering the coupling field from the mechanical system as

$$dW_m = -T_e d\theta_r \tag{2.110}$$

and the torque T_e as

$$T_e(\mathbf{i}, \theta_r) = \frac{\partial W_c(\mathbf{i}, \theta_r)}{\partial \theta_r} \tag{2.111}$$

Equation (2.110) must now be written in terms of θ_{rm}

$$dW_m = -T_e d\theta_{rm} \tag{2.112}$$

and since $\theta_{rm} = \frac{2}{P}\theta_r$, (2.111) becomes

$$T_e(\mathbf{i}, \theta_r) = \frac{P}{2}\frac{\partial W_c(\mathbf{i}, \theta_r)}{\partial \theta_r} \tag{2.113}$$

Analyzing a P-pole machine has turned out to be not as troublesome as we might have first thought. We are fortunate.

The torque and speed may be related by

$$T_e = J\left(\frac{2}{P}\right)\frac{d\omega_r}{dt} + B_m\left(\frac{2}{P}\right)\omega_r + T_L \tag{2.114}$$

where P is the number of poles, J is the inertia in kg m^2, B_m is the damping coefficient in N m s/rad, and T_L is positive for a load torque.

Example 2H ***Torque calculation for four-pole machine***
Consider the four-pole device shown in Figure 2.16. $L_{sr} = 0.1$ H, $N_s = 2N_r$, $L_{ls} = 2L_{lr}$ where $L_{ls} = 0.1L_{ms}$, $i_{as} = 2i_{ar} = 2$ A. Determine the coenergy W_c and the torque T_e.

From Section 2.2 with $\mathcal{R}_g$ the total reluctance of the two air gaps.

$$L_{sr} = \frac{N_s N_r}{\mathcal{R}_g} = 0.1 \text{ H} \tag{2H.1}$$

$$L_{ms} = \frac{N_s^2}{\mathcal{R}_g} \tag{2H.2}$$

$$L_{mr} = \frac{N_r^2}{\mathcal{R}_g} \tag{2H.3}$$

Therefore,

$$\begin{aligned} L_{ms} &= \frac{N_s}{N_r} L_{sr} \\ &= 2L_{sr} = 0.2 \text{ H} \end{aligned} \tag{2H.4}$$

and

$$\begin{aligned} L_{mr} &= L_{ms}\left(\frac{N_r}{N_s}\right)^2 \\ &= L_{ms}\left(\frac{1}{2}\right)^2 = 0.05 \text{ H} \end{aligned} \tag{2H.5}$$

Also

$$L_{ls} = 0.1\, L_{ms} = 0.02 \text{ H} \tag{2H.6}$$

$$L_{lr} = 0.5\, L_{ls} = 0.01 \text{ H} \tag{2H.7}$$

Now,

$$W_c = \frac{1}{2}L_{ss}i_{as}^2 + L_{asar}i_{as}i_{ar} + \frac{1}{2}L_{rr}i_{ar}^2 \tag{2H.8}$$

where

$$L_{ss} = L_{ls} + L_{ms} = 0.02 + 0.2 = 0.22\ \text{H} \tag{2H.9}$$

$$L_{asar} = 0.1\cos\theta_r \tag{2H.10}$$

$$L_{rr} = L_{lr} + L_{mr} = 0.01 + 0.05 = 0.06\ \text{H} \tag{2H.11}$$

$$\begin{aligned} W_c &= \left(\frac{1}{2}\right)(0.22)(2)^2 + (0.1)(2)(1)\cos\theta_r + \left(\frac{1}{2}\right)(0.06)(1)^2 \\ &= 0.44 + 0.2\cos\theta_r + 0.03 = 0.47 + 0.2\cos\theta_r\ \text{J} \end{aligned} \tag{2H.12}$$

The torque may be calculated using (2.113), that is

$$\begin{aligned} T_e(\mathbf{i}, \theta_r) &= \frac{P}{2}\frac{\partial W_c(\mathbf{i}, \theta_r)}{\partial \theta_r} \\ &= \left(\frac{4}{2}\right)\frac{\partial(0.2\cos\theta_r)}{\partial\theta_r} \\ &= -0.4\sin\theta_r\ \text{N m} \end{aligned} \tag{2H.13}$$

SP2.9.1 If, during steady-state operation, $\omega_r = \omega_e$, calculate the actual rotor speed in revolutions per minute of a 60-pole 60-Hz hydroturbine synchronous generator. [$\omega_{rm} = 120$ r/min]

SP2.9.2 Suppose $\theta_{esi}(0) = 45°$ in (2.103) and (2.104). For a six-pole two-phase stator, determine the location of the positive and negative maximum values of mmf$_s$ at time zero. [S^s at $\phi_s = 15°, 135°, 255°$; N^s at $\phi_s = 75°, 195°, 315°$]

2.10 Problems

2.1 Use the appropriate turns ratio to refer (2.1)–(2.4) to the 1 winding. Express v_1, v_2', λ_1, and λ_2' in terms of $i_1, i_2', L_{m1}, L_{12}', r_2'$, turns ratios, and θ_r.

2.2 Winding 4 is added to the stator of the device given in Figure 2.2 as shown in Figure 2.17. It is identical to winding 1 and displaced $\frac{\pi}{2}$ radians from it. Express $L_{44}, L_{14}, L_{24}, L_{34}$, and v_2. Let us mention in passing that the device in Figure 2.17 has the winding configuration of an induction motor. There are many induction motors serving you in your home: furnace fan, garbage disposal, air conditioner, etc.

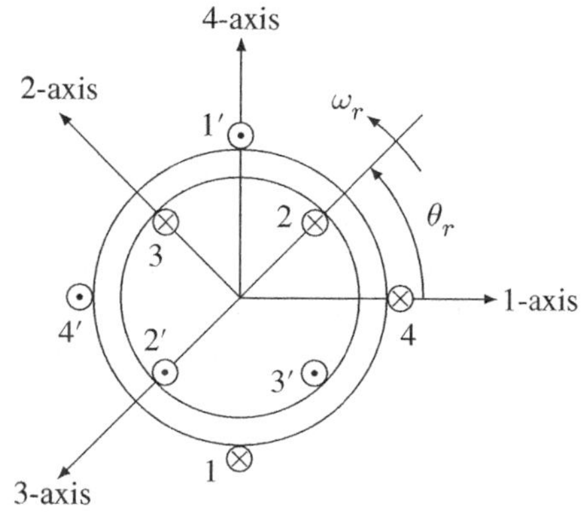

Figure 2.17 Two stator windings and two rotor windings.

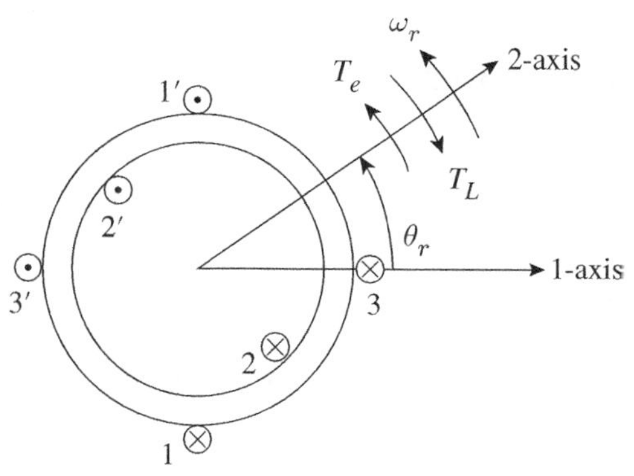

Figure 2.18 The stator windings and one rotor winding.

2.3 Use the device depicted in Figure 2.4 to show that the first two terms on the right-hand side of (2.31) sum to zero. That is, show that the coefficients of the *di* terms are equal.

2.4 Consider the device shown in Figure 2.18. The 3 winding which is identical to the 1 winding has been added to the stator of Figure 2.2. The device now has the winding configuration of an elementary synchronous machine. Express $T_e(i_1, i_2, i_3, \theta_r)$.

2.5 For the device shown in Figure 2.10, express the air gap mmf for balanced steady-state conditions as viewed by an observer running (a) clockwise at ω_e; (b) counterclockwise at $\frac{1}{2}\omega_e$ and (c) counterclockwise at $2\omega_e$. Let $\theta(0) = 0$.

2.6 Draw the phasor diagram showing the *qs* axis for Figure 2.11 with $\tilde{V}_{as}$ and $\tilde{I}_{as}$.

2.7 The expression for torque for a two-pole device is

$$T_e(\mathbf{i}, \theta_r) = \frac{\partial W_c(\mathbf{i}, \theta_r)}{\partial \theta_r}$$

where the coenergy is expressed in terms of currents and the electrical angular position. For balanced steady-state conditions the mechanical power output P_m is the shaft torque multiplied by the shaft angular velocity ω_{rm}. Derive an expression for P_m in terms of the electrical angular velocity ω_r and show that, for balanced steady-state conditions, it is valid for any P-pole device.

2.8 Consider the stator of the four-pole device shown in Figure 2.15. Assume these windings are concentrated with N_s total turns. Plot mmf$_s$ for $0 \leq \phi_s \leq 2\pi$.

References

1 Tesla, N. (2011). *My Inventions and Other Writings*. Penguin Books.

2 Park, R.H. (1929). Two-reaction theory of synchronous machines – generalized method of analysis – part I. *AIEE Trans.* 48: 716–727.

3 Krause, P.C., Wasynczuk, O., O'Connell, T.C., and Hasan, M. (2017). Tesla's contribution to electric machine analysis. *IEEE Trans. Energy Convers.* 32: 591–598.

4 White, D.C. and Woodson, H.H. (1959). *Electromechanical Energy Conversion*. Wiley.

5 Krause, P.C., Wasynczuk, O., Pekarek, S.D., and O'Connell, T.C. (2020). *Electromechanical Motion Devices*, 3e. IEEE Press/Wiley.

6 Sudhoff, S.D. (2013). *Power Magnetic Devices*. IEEE Press/Wiley.

3

Electric Machines

3.1 Introduction

This chapter is devoted to a preliminary treatment of the direct current (dc), the permanent-magnetic alternating current (ac), and the induction machines. This is not an in-depth analysis, but rather the basic analysis sufficient to incorporate these machines into the basic drive and power areas. We have set forth the basic analysis methods in Chapter 2, and we will assume that the student has studied these concepts and is able to work with them. Please understand that if the student wishes to work in the drives or power areas, this analysis of these machines must be supplemented with a more in-depth treatment [1–4]. Reference frame theory is introduced and used in the derivation of the dynamic and steady-state equations. Although necessary to this derivation, it can be deemphasized as is pointed out in the text.

3.2 Direct-current Machine

The dc machine is not as widely used today as it was in the past. The dc generator has been replaced by solid-state rectifiers which convert alternating current into direct current with provisions to control the magnitude of the dc voltage and, in drive applications, the dc motor is being replaced by the voltage controlled permanent-magnet synchronous machine (brushless dc drive) and/or the field-oriented induction motor. Nevertheless, it is still desirable to devote some time to the dc machine in an introductory course since it is still used as a low-power drive motor, especially in automotive applications. There is another and perhaps more important reason to consider the dc machine. Although maintenance and environmental issues hamper the use of dc machines, this device is the only electric machine that is designed with the stator and rotor

Introduction to the Analysis of Electromechanical Systems, First Edition.
Paul C. Krause, Oleg Wasynczuk, and Timothy O'Connell.

magnetomotive forces (mmfs) orthogonal, which inherently produces maximum torque per ampere. With the advent of electronic switching devices, there has become a huge effort to control the permanent-magnet ac synchronous and induction machines so as to emulate the performance characteristics of the dc motor without the maintenance and environmental problems inherent to the dc machines. This section is an attempt to treat dc machines sufficient to introduce the reader to the operating principles of dc machines with focus on the shunt-connected and permanent-magnet dc machines, thereby setting the stage for a comparison of the operating characteristics with the voltage controlled permanent-magnet ac drive and the field-oriented induction motor.

3.2.1 Commutation

An elementary dc machine is shown in cross section in Figures 3.1 and 3.2. The field winding is carrying a direct current i_f into the paper at f_1 out at f_1' and then to f_2 and out at f_2'. A voltage v_f is applied across f_1 and f_2'. With positive i_f, the field winding creates an mmf that is stationary and positive in the f axis.

The armature or rotor consists of two parallel windings: the a winding and the A winding. Each winding has four coils with each coil connected to two segments of the commutator. The commutator is fixed to the rotor and the brushes, which are graphite, riding on the commutator connect the rotating circuits (a winding and A winding) to the stationary voltage v_a, which could be connected to a dc source or to a load if the device is operating as a generator. Note that in Figure 3.1 the dc current i_a is flowing into the top brush which is straddling two of the eight segments of the commutator. Each segment is insulated from the others. The top brush is short-circuiting the A_4 coil; the bottom brush is short-circuiting the a_4 coil. In Figure 3.2, the brushes are not commutating any windings. Sinsusoidal voltages are induced in each of the coils due to the constant field current producing a stationary mmf$_s$ and the winding rotating in this constant mmf$_s$. Due to the action of the commutator, the mmf$_r$ (a axis) of the rotor is also essentially stationary and orthogonal to the field mmf$_s$ (f axis).

The full-wave rectified voltages in Figures 3.1 and 3.2 are the open-circuit voltages of one parallel path between brushes. This is referred to as the back voltage. It is the induced voltage that exists only when the rotor is turning. In Figure 3.1 and 3.2, the parallel windings consist of four coils and produce an mmf that is orthogonal to the mmf produced by the stator winding, the f winding. We'll see the importance of this orthogonal positioning of the magnetic axes when we derive the expression of the torque.

It is important to understand that the commutator and brush combination is to change the direction of current flow in the rotor windings so that for positive

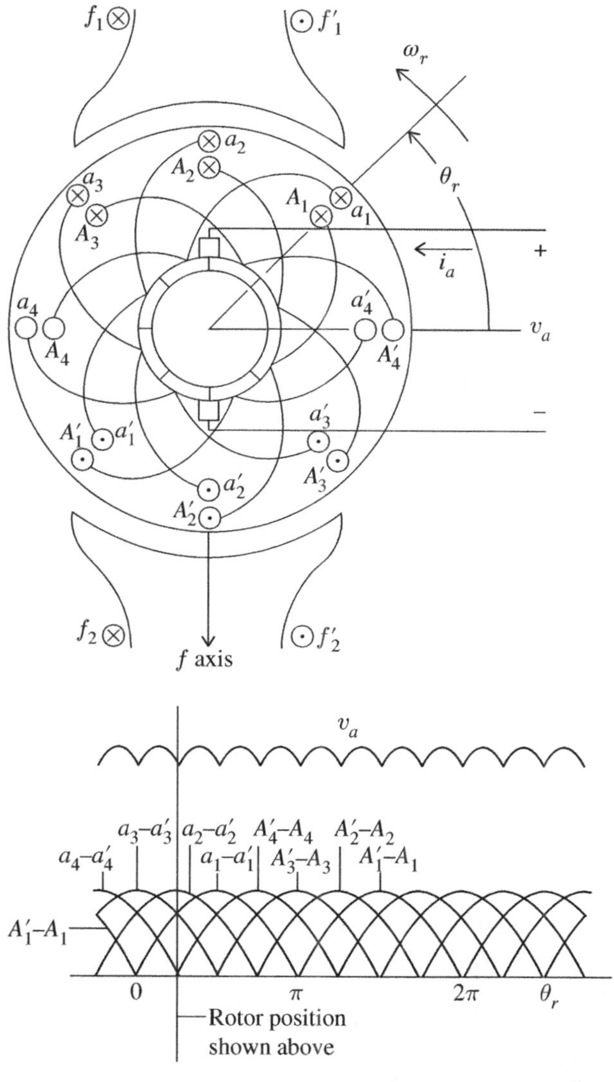

Figure 3.1 A dc machine with parallel armature windings.

current into the machine, it flows into the paper over the top one-half of the rotor and out over the bottom one-half for motor action. The current is reversed for generator action.

A disassembled two-pole 0.1-hp 6-V 12 000-r/min permanent-magnet dc motor is shown in Figure 3.3. The magnets, which replace the stator field winding, are samarium cobalt and the device is used to drive hand-held battery-operated

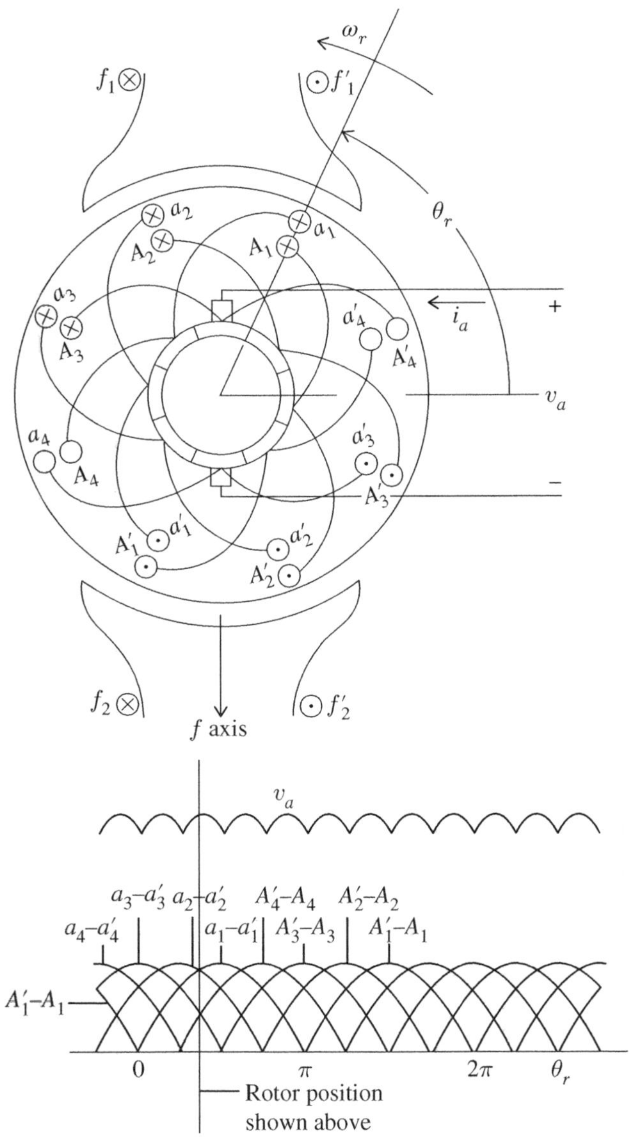

Figure 3.2 Same as Figure 3.1 with rotor advanced approximately 22.5° counterclockwise.

Figure 3.3 Two-pole 0.1-hp 6-V 12 000-r/min permanent-magnet dc motor. Source: Courtesy of Vick ElectroMech.

surgical instruments. Although some of these terms are new to us, they will be defined as we go along.

3.2.2 Voltage and Torque Equations

It is advantageous to first consider the dc machine with a field and armature winding before turning to the permanent-magnet device exclusively. Although rigorous derivation of the voltage and torque equations is possible, it is rather lengthy and little is gained since these relationships may be deduced. The armature coils revolve in a magnetic field established by a current flowing in the field winding. We have established that voltage is induced in these coils by virtue of this rotation. However, the action of the commutator causes the armature coils to appear as a stationary winding with its magnetic axis orthogonal to the magnetic axis of the field winding. In other words, the stator and rotor mmfs are orthogonal. Consequently, voltages are not induced in one winding due to the time rate of change of the current flowing in the other (transformer action). Mindful of these conditions, we can write the field and armature voltage equations in matrix form as [5]

$$\begin{bmatrix} v_f \\ v_a \end{bmatrix} = \begin{bmatrix} r_f + pL_{FF} & 0 \\ \omega_r L_{AF} & r_a + pL_{AA} \end{bmatrix} \begin{bmatrix} i_f \\ i_a \end{bmatrix} \tag{3.1}$$

where L_{FF} and L_{AA} are the self-inductances of the field and armature windings, respectively, and p is the short-hand notation for the operator d/dt. The rotor speed is denoted as ω_r, and L_{AF} is the mutual inductance between the field and the rotating armature coils, which is readily determined from the open-circuited voltage. The above equation suggests the equivalent circuit shown in Figure 3.4. The voltage induced in the armature circuit, $\omega_r L_{AF} i_f$, is commonly referred to as the counter or back electromotive force (emf). It also represents the open-circuit armature voltage from which L_{AF} can be readily determined. The equivalent circuit

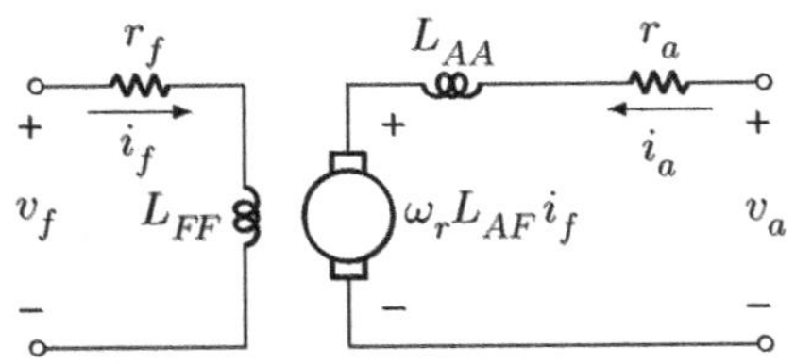

Figure 3.4 Equivalent circuit of dc machine.

shown in Figure 3.4 is for a separately excited machine where v_f is from a separate dc source. When $v_f = v_a$ it is a shunt machine.

A substitute variable often used is

$$k_v = L_{AF} i_f \tag{3.2}$$

We will find that this substitute variable is particularly convenient and frequently used. Even though a permanent-magnet dc machine has no field circuit, the constant field flux produced by the permanent magnet is analogous to a dc machine with a constant k_v.

The torque can be determined by first expressing mmf_s and mmf_r. From Figures 3.1 and 3.2,

$$\text{mmf}_s = \frac{N_f}{2} \sin \phi_s i_f \tag{3.3}$$

$$\text{mmf}_r = \frac{N_a}{2} \cos \phi_s i_a \tag{3.4}$$

The mmf's are orthogonal. The field flux Φ_f may be expressed as $\frac{\text{mmf}_s}{\mathcal{R}}$ where $\mathcal{R}$ is an equivalent reluctance. The torque may be expressed by the cross product of the maximum mmf_a and Φ_f. Thus

$$\begin{aligned} T_e &= \left(\frac{N_f i_f}{\mathcal{R}} \right) (N_a i_a) \\ &= \frac{N_f N_a}{\mathcal{R}} i_f i_a \\ &= L_{AF} i_f i_a \end{aligned} \tag{3.5}$$

The torque and rotor speed are related by

$$T_e = J \frac{d\omega_r}{dt} + B_m \omega_r + T_L \tag{3.6}$$

where J is the inertia of the rotor and rigidly connected mechanical load. The units of the inertia are kg m^2 or J s^2. A positive electromagnetic torque T_e acts to turn the rotor in the direction of increasing θ_r. The load torque T_L is positive for a torque, on the shaft of the rotor, which opposes a positive electromagnetic torque T_e. The constant B_m is a damping coefficient associated with the mechanical rotational system of the machine. It has the units of N m s/rad, and it is generally small and often neglected.

Although we will focus on the permanent-magnet dc motor, it is worthwhile to take a moment to mention that we have established the basis for several types of dc machines. In particular, the machine shown in Figure 3.4 is a separately excited dc machine. If we connect the field winding in parallel with the armature winding, it becomes a shunt-connected dc machine. If the field winding is connected in series with the armature winding, it is a series-connected dc machine. If two windings are used, one in parallel with and another in series with the armature it is referred to as a compound-connected dc machine. Clearly, this is an overly simplistic description, and the reader is referred to [1] for a more detailed consideration of these machine types.

Before proceeding, it is appropriate to mention briefly generator action even though the dc generator has been replaced by the ac to dc converter. We see from Figure 3.4 that if $\omega_r L_{AF} i_f$ is greater than v_a, i_a is reversed and we have generator action. In this case, T_L is negative since the dc machine is being driven and T_e (3.5) is negative and, in the steady state from (3.6), $\frac{d\omega_r}{dt}$ is zero and B_m is small.

3.2.3 Permanent-magnet dc Machine

In the case of the permanent-magnet dc machine, $L_{AF} I_f$ is replaced with k_v whereupon the steady-state armature voltage equation becomes

$$V_a = r_a I_a + k_v \omega_r \tag{3.7}$$

If (3.7) is solved for I_a and substituted in to (3.5) with $L_{AF} I_f$ replaced by k_v, the steady-state torque may be expressed

$$\begin{aligned} T_e &= k_v I_a \\ &= \frac{k_v V_a - k_v^2 \omega_r}{r_a} \end{aligned} \tag{3.8}$$

The steady-state torque-speed characteristic is shown in Figure 3.5.

It is apparent from Figure 3.5 that the stall ($\omega_r = 0$) torque could be made larger for a given armature voltage by reducing r_a. Although the machine may be designed with a smaller armature resistance, there is a problem since, at stall, the steady-state armature current is limited by the armature resistance; hence, for a constant V_a, reducing r_a will result in a larger I_a which can cause damage to the brushes. On the other hand, increasing the starting torque by reducing r_a causes the torque-speed characteristics to have a steeper slope which results in a smaller change in speed for a given change in load torque during normal (near rated) operation. If, however, the armature voltage is reduced during the starting period to protect the brushes, the desirable characteristic of a small speed

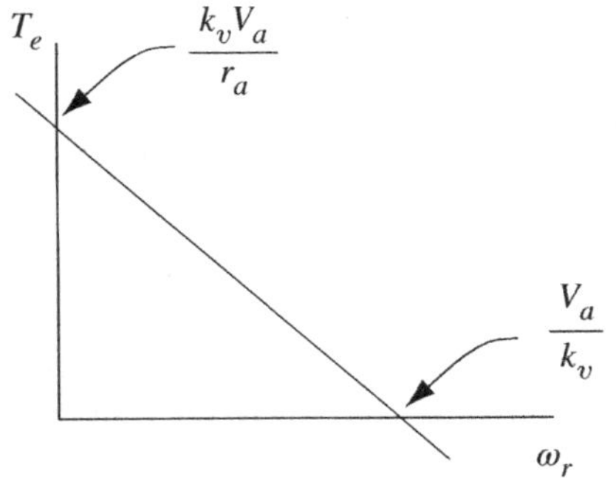

Figure 3.5 Steady-state torque-speed characteristic of a permanent-magnet dc machine.

change during load torque variations during normal operation could be achieved. In fact, controlled regulation of the armature voltage is generally employed for large horsepower machines by using a converter; however, low-power permanent-magnet dc machines are generally supplied from a battery as in the case of the automobile and, therefore, a large armature resistance is necessary to prevent brush damage during the early part of the starting period. Fortunately, a small speed variation during load torque changes is not required in many applications of the permanent-magnet dc machine; therefore, the steep torque-speed characteristics are not necessary. We will consider the dynamic performance of a permanent-magnet dc motor in the following Example 3A. In the next section, we will discover that the brushless dc machine can be controlled to have torque-speed characteristics similar to that of the permanent-magnet dc motor.

Example 3A ***Calculating machine parameters***

A permanent-magnet dc motor similar to that shown in Figure 3.3 is rated at 6 V with the following parameters: $r_a = 7\,\Omega$, $L_{AA} = 120$ mH, $k_T = 2$ oz in/A, and $J = 150\ \mu$oz in s^2. (a) Determine the stall torque and the no-load speed. (b) A torque load of 0.5 oz· in is applied, determine the steady-state ω_r. (c) Determine the efficiency at this load.

First, let us convert k_T and J to units that we have been using in this text. In this regard, we will convert the inertia to kg m^2 which is the same as N m s^2. To do this, we must convert ounces to Newtons and inches to meters (Appendix A). Thus,

$$J = \frac{150 \times 10^{-6}}{(3.6)(39.37)} = 1.06 \times 10^{-6}\ \text{kg m}^2 \tag{3A.1}$$

We have not seen k_T before. It is the torque constant and, if expressed in the appropriate units, it is numerically equal to k_v. When k_v is used in the expression for $T_e (T_e = k_v i_a)$, it is often referred to as the *torque constant* and denoted k_T. When used in the voltage equation, it is always denoted k_v. Now we must convert oz in to N m, whereupon k_T equals our k_v; hence,

$$k_v = \frac{2}{(3.6)(39.37)} = 1.41 \times 10^{-2}\text{N m/A} = 1.41 \times 10^{-2}\text{ V s/rad} \tag{3A.2}$$

(a) The stall torque is

$$T_e = \frac{k_v V_a}{r_a} = \frac{(1.41 \times 10^{-2})(6)}{7} = 1.21 \times 10^{-2}\text{ N m} \tag{3A.3}$$

The no-load speed is

$$\omega_r = \frac{V_a}{k_v} = \frac{6}{1.41 \times 10^{-2}} = 425.5\text{ rad/s} \tag{3A.4}$$

(b) $$T_L = \frac{0.5}{(3.6)(39.37)} = 3.5 \times 10^{-3}\text{ N m} \tag{3A.5}$$

From (3.8) with $T_e = T_L$

$$\omega_r = \frac{V_a}{k_v} - \frac{T_L V_a}{k_v^2} = \frac{6}{1.41 \times 10^{-2}} - \frac{(3.5 \times 10^{-3})(7)}{(1.41 \times 10^{-2})^2}$$
$$= 425.5 - 123.2 = 302.3\text{ rad/s} \tag{3A.6}$$

(c) $$T_e = k_v I_a$$
$$3.5 \times 10^{-3} = (1.41 \times 10^{-2}) I_a$$
$$I_a = \frac{3.5 \times 10^{-3}}{1.41 \times 10^{-2}} = 0.248\text{ A} \tag{3A.7}$$

$$P_{\text{loss}} = r_a I_a^2 = (7)(0.248)^2 = 0.431\text{ W} \tag{3A.8}$$

$$P_{\text{in}} = V_a I_a = (6)(0.248) = 1.488\text{ W} \tag{3A.9}$$

$$P_{\text{out}} = P_{\text{in}} - P_{\text{loss}} = 1.488 - 0.437 = 1.057\text{ W} \tag{3A.10}$$

$$\text{eff} = \frac{P_{\text{out}}}{P_{\text{in}}} \times 100 = \frac{1.057}{1.488} \times 100 = 71\% \tag{3A.11}$$

Note that

$$P_{\text{out}} = T_e \omega_r = 3.5 \times 10^{-3} \times 302.3 = 1.058\text{ W} \tag{3A.12}$$

which is essentially equal to (3A.10).

SP3.2.1 The peak value of the voltage induced in one coil shown in Figure 3.1 is 1 V. Determine, from Figure 3.1, the maximum and minimum value of v_a. [2.613 V; 2.414 V]

SP3.2.2 When a 12-V permanent-magnet dc motor is driven at 100 rad/s, the open-circuit voltage is 10 V. Calculate k_v. [$k_v = 0.1$ V s/rad]

SP3.2.3 The armature applied voltage is 240 V; the rotor speed is constant at 50 rad/s and $I_a = 15$ A. The armature resistance is 1 Ω and $L_{AF} = 1$ H. Calculate the steady-state field current. [$I_f = 4.5$ A]

SP3.2.4 Multiply the expression for v_a given in (3.1) by i_a and identify all terms. [$v_a i_a$ – input power to armature, $i_a^2 r_a$ – armature ohmic loss, $L_{AA} i_a p i_a$ – change of energy stored in L_{AA}, $L_{AF} i_f i_a \omega_r = T_e \omega_r$ = output power]

SP3.2.5 Is SP3.2.4 an alternate method of deriving an expression for torque? [Yes; $L_{AF} i_f i_a$ is the coefficient of $p\theta_r$ or ω_r]

SP3.2.6 A 12-V permanent-magnet dc motor has an armature resistance of 12 Ω and $k_v = 0.01$ V s/rad. Calculate the steady-state stall torque (T_e with $\omega_r = 0$). [$T_e = 0.01$ N m]

3.3 Permanent-magnet ac Machine

The permanent-magnet ac machine is used in many low-power drive application; it has emerged as the electric machine of choice. Moreover, it provides an excellent example of power-electronic control when operated as a brushless dc motor, and the analysis of the permanent-magnet ac machine is similar in many respects to that of the synchronous machine, which we will cover in Chapter 6.

3.3.1 Two-phase Permanent-magnet ac Machine

A two-pole two-phase permanent-magnet ac machine is shown in Figure 3.6. The stator windings are identical, sinusoidally distributed windings each with N_s equivalent turns and resistance, r_s. The magnetic axes of the stator windings are denoted the *as* and the *bs* axis. The *d* axis (direct axis) is fixed at the center of the magnetic flux that enters the air gap from the permanent-magnet rotor and the *q* axis (quadrature axis) is displaced $\frac{1}{2}\pi$ counterclockwise (ccw) from the *d* axis. These are the same *q* axis and *d* axis we introduced in Chapter 2. The angular position about the stator is denoted ϕ_s, referenced to the *as* axis. The angular position about the rotor is ϕ_r, referenced to the *q* axis. The angular velocity of the rotor is ω_r and θ_r is the angular displacement of the rotor measured from the *as* axis to the *q* axis. Thus, a given point on the rotor surface at the angular position ϕ_r may be related to an adjacent point on the inside stator surface with angular position ϕ_s as

$$\phi_s = \theta_r + \phi_r \tag{3.9}$$

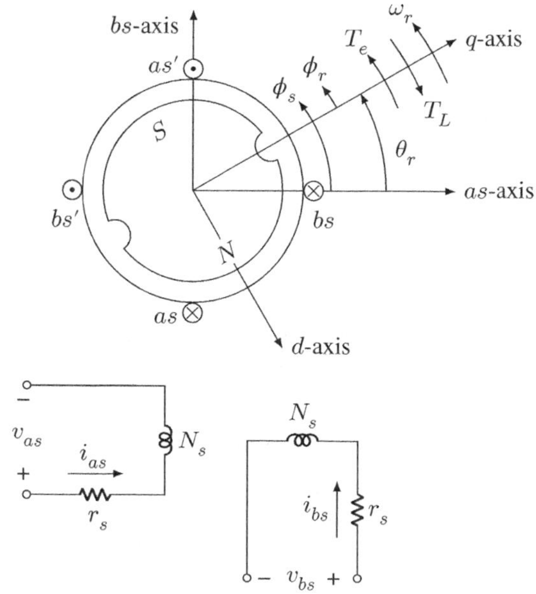

Figure 3.6 Two-pole two-phase permanent-magnet ac machine.

The electromechanical torque T_e and the load torque T_L are also indicated in Figure 3.6. The torque T_e is assumed positive in the direction of increasing θ_r; T_L is positive in the opposite direction. Electromagnetic torque is produced by the interaction of the poles of the permanent-magnet rotor and the poles resulting from the rotating air-gap mmf established by balanced currents flowing in the stator windings given by (2.46).

A four-pole three-phase 28-V $\frac{1}{3}$-hp permanent-magnet ac machine is shown in Figure 3.7. The disassembled motor is shown in Figure 3.7a, wherein the stator windings are visible. The opposite end of the stator housing is shown in Figure 3.7b. Housed therein are Hall-effect sensors, which are used to determine the position of the permanent magnet, the drive inverter, the filter capacitor, and the logic circuitry. We'll talk more about the sensors and inverter in Chapters 4 and 5.

The voltage equations for the two-pole two-phase permanent-magnet ac machine shown in Figure 3.6 may be expressed as

$$v_{as} = r_s i_{as} + \frac{d\lambda_{as}}{dt} \tag{3.10}$$

$$v_{bs} = r_s i_{bs} + \frac{d\lambda_{bs}}{dt} \tag{3.11}$$

In matrix form,

$$\mathbf{v}_{abs} = \mathbf{r}_s \mathbf{i}_{abs} + p\boldsymbol{\lambda}_{abs} \tag{3.12}$$

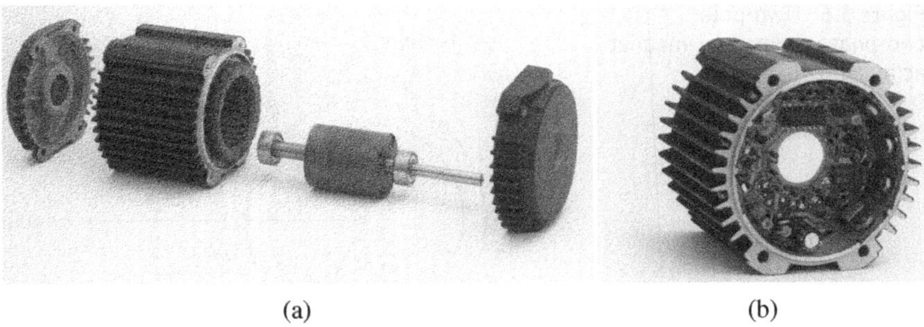

Figure 3.7 Four-pole three-phase 28-V $\frac{1}{3}$-hp permanent-magnet ac machine. (a) Disassembled motor; (b) stator housing showing Hall-effect sensors. Source: Courtesy of EG and G Rotron.

where p is the operator d/dt, and for voltages, currents, and flux linkages

$$(\mathbf{f}_{abs})^T = [f_{as}\, f_{bs}] \tag{3.13}$$

with

$$\mathbf{r}_s = \begin{bmatrix} r_s & 0 \\ 0 & r_s \end{bmatrix} \tag{3.14}$$

The flux linkage equations may be expressed as

$$\lambda_{as} = L_{asas} i_{as} + L_{asbs} i_{bs} + \lambda_{asm} \tag{3.15}$$

$$\lambda_{bs} = L_{bsas} i_{as} + L_{bsbs} i_{bs} + \lambda_{bsm} \tag{3.16}$$

In matrix form,

$$\boldsymbol{\lambda}_{abs} = \mathbf{L}_s \mathbf{i}_{abs} + \boldsymbol{\lambda}'_m \tag{3.17}$$

where $\boldsymbol{\lambda}'_m$ is the column vector

$$\boldsymbol{\lambda}'_m = \begin{bmatrix} \lambda_{asm} \\ \lambda_{bsm} \end{bmatrix} = \lambda'_m \begin{bmatrix} \sin\theta_r \\ -\cos\theta_r \end{bmatrix} \tag{3.18}$$

In (3.18), λ'_m is the amplitude of the flux linkages established by the permanent magnet as viewed from the stator phase windings. In other words, the magnitude of λ'_m is proportional to the magnitude of the open-circuit sinusoidal voltage induced in each stator phase winding due to rotation of the rotor. The rotor displacement θ_r is expressed as

$$\frac{d\theta_r}{dt} = \omega_r \tag{3.19}$$

We will assume that the reluctance of the rotor of the permanent-magnet ac machine is the same in the q and d axes. Actually, this assumption may be an oversimplification in some cases; however, it markedly reduces our work and allows us to establish the operating principles of the permanent-magnet ac machine without significant error.

With the assumption of a uniform air gap, the mutual inductance between the *as* and *bs* windings is zero. Since the windings are identical, the self-inductances L_{asas} and L_{bsbs} are equal and denoted as L_{ss}. As in the case of a transformer, the self-inductance is made up of a leakage L_{ls} and a magnetizing inductance L_{ms}. Thus,

$$L_{ss} = L_{ls} + L_{ms} \tag{3.20}$$

The machine is designed to minimize the leakage inductance, which generally makes up approximately 10% of L_{ss}. Since the air-gap is assumed to be uniform, the magnetizing inductances of the *as* and *bs* windings are constant and equal; therefore, the self-inductance matrix L_s may be expressed as

$$L_s = \begin{bmatrix} L_{ss} & 0 \\ 0 & L_{ss} \end{bmatrix} \tag{3.21}$$

We can derive an expression for torque T_e by first expressing the coenergy for a magnetically linear system.

$$\begin{aligned} W_c &= W_f \\ &= \frac{1}{2}L_{asas}i_{as}^2 + \frac{1}{2}L_{bsbs}i_{bs}^2 + W_{pm} + \lambda_{asm}i_{as} + \lambda_{bsm}i_{bs} \end{aligned} \tag{3.22}$$

where W_{pm} is the energy associated with the permanent magnet which is assumed to be constant and λ_{asm} and λ_{bsm} are given by (3.18). These flux linkages may be thought of as a mutual inductance times a current corresponding to the strength of the permanent magnet. Taking the partial derivative with respect to θ_r yields

$$\begin{aligned} T_e &= \frac{P}{2}\frac{\partial W_c}{\partial \theta_r} \\ &= \frac{P}{2}\lambda_m'(i_{as}\cos\theta_r + i_{bs}\sin\theta_r) \end{aligned} \tag{3.23}$$

The above expression is positive for motor action. The torque and rotor speed may be related as

$$T_e = J\left(\frac{2}{P}\right)\frac{d\omega_r}{dt} + B_m\left(\frac{2}{P}\right)\omega_r + T_L \tag{3.24}$$

where J is in kg m^2; it is the inertia of the rotor and rigidly connected mechanical load. Since we will be concerned primarily with motor action, the load torque T_L is assumed positive, as indicated in Figure 3.6. The constant B_m is a damping coefficient associated with the rotational system of the machine and mechanical load. It has the units N m s/rad, and it is generally small and often neglected in the case of the machine but may be considerable for the mechanical load.

Example 3B ***Measuring λ_m' of a permanent-magnet machine***

The parameters of a four-pole two-phase permanent-magnet ac machine are $r_s = 3.4\ \Omega$, $L_{ls} = 1.1$ mH, and $L_{ms} = 11$ mH. When the device is driven at 1000 r/min, the open-circuit phase voltage is sinusoidal with a peak-to-peak value of 34.6 V. Determine λ_m'. The actual rotor speed at which the measurement was taken is

$$\omega_{rm} = \frac{(\text{r}/\min)(\text{rad}/\text{r})}{\text{s}/\min}$$

$$= \frac{(1000)(2\pi)}{60} = \frac{100}{3}\pi \text{ rad/s} \tag{3B.1}$$

From (2.109), the electrical angular velocity is

$$\omega_r = \frac{P}{2}\omega_{rm}$$

$$= \frac{4}{2}\frac{100\pi}{3} = \frac{200}{3}\pi \text{ rad/s} \tag{3B.2}$$

With the phases open-circuited, $i_{as} = i_{bs} = 0$. Thus, from (3.10) and (3.18),

$$v_{as} = \frac{d(\lambda_m' \sin\theta_r)}{dt} = \lambda_m' \omega_r \cos\theta_r \tag{3B.3}$$

Now the peak-to-peak voltage is 34.6 V; hence, from (3B.3), with the peak-to-peak voltage divided by 2, we have

$$\frac{34.6}{2} = \lambda_m' \left(\frac{200}{3}\pi\right) \tag{3B.4}$$

Solving for λ_m' yields

$$\lambda_m' = \frac{(34.6)(3)}{(2)(200\pi)} = 0.0826 \text{ V s/rad} \tag{3B.5}$$

3.3.2 Reference Frame Analysis of a Permanent-magnet ac Machine

This reference frame derivation may be deemphasized and in an in-depth study resumed with the subsection on steady-state analysis. The permanent-magnet ac machine is unsymmetrical in the rotor due to the permanent magnet only in one axis. Therefore, we must use the rotor reference frame. The voltage equations for the stator equations in the rotor reference frame are given by (2.90) and (2.91). These equations are repeated here

$$v_{qs}^r = r_s i_{qs}^r + \omega_r \lambda_{ds}^r + p\lambda_{qs}^r \tag{3.25}$$

$$v_{ds}^r = r_s i_{ds}^r - \omega_r \lambda_{qs}^r + p\lambda_{ds}^r \tag{3.26}$$

We need to determine λ_{qs}^r and λ_{ds}^r; we can get there from (3.17) which is

$$\boldsymbol{\lambda}_{abs} = \mathbf{L}_s \mathbf{i}_{abs} + \boldsymbol{\lambda}_m' \tag{3.27}$$

where

$$\mathbf{L}_s = \begin{bmatrix} L_{ss} & 0 \\ 0 & L_{ss} \end{bmatrix} \tag{3.28}$$

and

$$L_{ss} = L_{ls} + L_{ms} \tag{3.29}$$

also

$$\boldsymbol{\lambda}'_m = \lambda'_m \begin{bmatrix} \sin\theta_r \\ -\cos\theta_r \end{bmatrix} \tag{3.30}$$

For a magnetically linear system, the stator flux linkages are expressed by (3.27). Substituting the change of variables into (3.27) yields

$$(\mathbf{K}_s^r)^{-1}\boldsymbol{\lambda}_{qds}^r = \mathbf{L}_s(\mathbf{K}_s^r)^{-1}\mathbf{i}_{qds}^r + \boldsymbol{\lambda}'_m \tag{3.31}$$

Pre-multiplying by $\mathbf{K}_s^r$ and substituting (3.28) and (3.29) for $\mathbf{L}_s$ yields

$$\boldsymbol{\lambda}_{qds}^r = \begin{bmatrix} L_{ls}+L_{ms} & 0 \\ 0 & L_{ls}+L_{ms} \end{bmatrix} \begin{bmatrix} i_{qs}^r \\ i_{ds}^r \end{bmatrix} + \lambda'^r_m \begin{bmatrix} 0 \\ 1 \end{bmatrix} \tag{3.32}$$

Make sure you agree that $\mathbf{K}_s^r\mathbf{L}_s(\mathbf{K}_s^r)^{-1}$ is $\mathbf{L}_s$. To be consistent with our previous notation, we have added the superscript r to λ'_m, thus $\lambda'_m = \lambda'^r_m$. We see from (3.32) that, in our new system of variables, the flux linkage created by the rotating permanent magnet appears constant. Hence, the $_{qs}^r$ and $_{ds}^r$ circuits are fixed relative to the permanent magnet and therefore fixed with the rotor speed ω_r. We have accomplished the goal of eliminating flux linkages that vary with θ_r. One more thing before we leave (3.32), please note that the stator leakage inductance is contained in λ_{qs}^r and λ_{ds}^r. This tells us that we must be careful where we conceptualize the location of these windings. If we consider them to be fixed on the rotor, there would not be a stator leakage inductance. Therefore, we have little choice but to think of these windings "rotating through iron"; however, it is common to refer to the $_{qs}^r$ and $_{ds}^r$ windings as "fixed in the rotor."

In expanded form, the voltage equations are

$$v_{qs}^r = r_s i_{qs}^r + \omega_r \lambda_{ds}^r + p\lambda_{qs}^r \tag{3.33}$$

$$v_{ds}^r = r_s i_{ds}^r - \omega_r \lambda_{qs}^r + p\lambda_{ds}^r \tag{3.34}$$

where

$$\lambda_{qs}^r = L_{ss} i_{qs}^r \tag{3.35}$$

$$\lambda_{ds}^r = L_{ss} i_{ds}^r + \lambda'^r_m \tag{3.36}$$

Equations (3.33)–(3.36) suggest the $_{qs}^r$- and $_{ds}^r$- equivalent machine and the equivalent circuits shown in Figure 3.8.

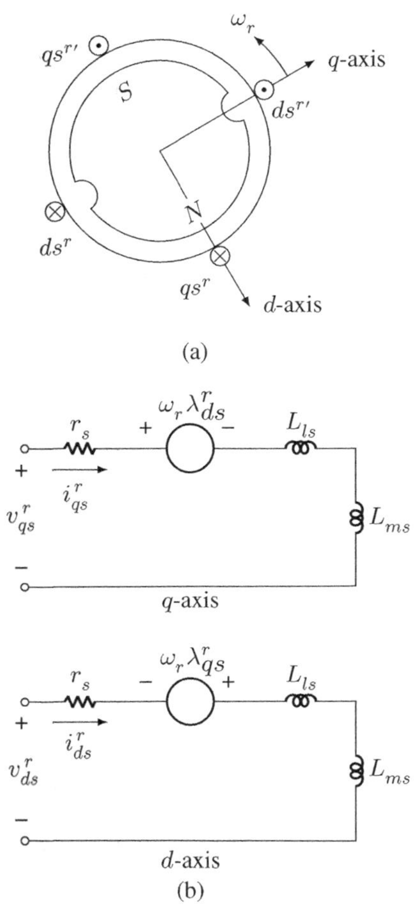

Figure 3.8 (a) The q_s^r- and d_s^r-equivalent permanent-magnet ac machine, (b) quadrature- and direct-axis equivalent circuits.

Substituting (3.35) and (3.36) into (3.33) and (3.34) and, since λ'^r_m is constant, $p\lambda'^r_m = 0$ we have

$$v^r_{qs} = (r_s + pL_{ss})i^r_{qs} + \omega_r L_{ss} i^r_{ds} + \omega_r \lambda'^r_m \tag{3.37}$$

$$v^r_{ds} = (r_s + pL_{ss})i^r_{ds} - \omega_r L_{ss} i^r_{qs} \tag{3.38}$$

The above equations may be rewritten in matrix form as

$$\begin{bmatrix} v^r_{qs} \\ v^r_{ds} \end{bmatrix} = \begin{bmatrix} r_s + pL_{ss} & \omega_r L_{ss} \\ -\omega_r L_{ss} & r_s + pL_{ss} \end{bmatrix} \begin{bmatrix} i^r_{qs} \\ i^r_{ds} \end{bmatrix} + \begin{bmatrix} \omega_r \lambda'^r_m \\ 0 \end{bmatrix} \tag{3.39}$$

The expression for electromagnetic torque positive for motor action is given by (3.23) as

$$T_e = \frac{P}{2} \lambda'^r_m (i_{as} \cos\theta_r + i_{bs} \sin\theta_r) \tag{3.40}$$

Since the bracketed terms are i^r_{qs}, the torque can now be expressed as

$$T_e = \frac{P}{2}\lambda'^r_m i^r_{qs} \tag{3.41}$$

Show that the torque may also be expressed

$$T_e = \frac{P}{2}(\lambda^r_{ds} i^r_{qs} - \lambda^r_{qs} i^r_{ds}) \tag{3.42}$$

The expressions for torque given by (3.40)–(3.42) are valid in general; unbalanced, magnetically nonlinear, and transient conditions.

Let us now assume that the applied stator voltages are

$$v_{as} = v_s \cos\theta_{esv} \tag{3.43}$$

$$v_{bs} = v_s \sin\theta_{esv} \tag{3.44}$$

where v_s is the amplitude of the stator applied voltages and

$$\omega_e = \frac{d\theta_{esv}}{dt} \tag{3.45}$$

Transforming these voltages to the rotor reference frame yields

$$v^r_{qs} = v_s \cos\phi_v \tag{3.46}$$

$$v^r_{ds} = -v_s \sin\phi_v \tag{3.47}$$

where

$$\phi_v = \theta_{esv} - \theta_r \tag{3.48}$$

We will find that if ϕ_v is zero, the device has the operating characteristics (T_e versus ω_r) similar to that of a permanent-magnet dc machine (see Figure 5.11); this is the reason a permanent-magnet ac machine which has no brushes or commutator is called a "brushless dc machine."

When the permanent-magnet ac machine is operated as a brushless dc drive, the converter is controlled to reduce the frequency of the applied voltage when the machine is loaded and increase the frequency when the load is decreased. It is a variable frequency drive system, and in addition the converter maintains ϕ_v at zero which means that the converter controls the maximum value of v_{as} to be in phase with the q axis. Although $\phi_v = 0$ is the common mode of operation, it can be advanced which puts the maximum voltage ahead of the q axis to give maximum torque per voltage or maximum torque per ampere. Please understand that with $\omega_r = \omega_e$ we could use either r or e as the superscript on the substitute variables. We'll use r to call attention to the fact that the control makes $\omega_e = \omega_r$.

3.3.3 Three-phase Permanent-magnet ac Machine

Due to the $\frac{2}{3}$ factor in Park's transformation, the v^r_{qs} and v^r_{ds} equations are identical in form to those for the two-phase given by (3.25) and (3.26). We need only to multiply the expression for torque, T_e, by $\frac{3}{2}$; however, due to the mutual coupling between stator phases the flux linkages become

$$\boldsymbol{\lambda}_{abcs} = \begin{bmatrix} L_{ls} + L_{ms} & -\frac{1}{2}L_{ms} & -\frac{1}{2}L_{ms} \\ -\frac{1}{2}L_{ms} & L_{ls} + L_{ms} & -\frac{1}{2}L_{ms} \\ -\frac{1}{2}L_{ms} & -\frac{1}{2}L_{ms} & L_{ls} + L_{ms} \end{bmatrix} \begin{bmatrix} i_{as} \\ i_{bs} \\ i_{cs} \end{bmatrix} + \lambda'_m \begin{bmatrix} \sin\theta_r \\ \sin\left(\theta_r - \frac{2}{3}\pi\right) \\ \sin\left(\theta_r + \frac{2}{3}\pi\right) \end{bmatrix} \tag{3.49}$$

Since $i_{as} + i_{bs} + i_{cs} = 0$ for the three-wire system, (3.49) may also be written

$$\boldsymbol{\lambda}_{abcs} = \mathbf{L}_s \mathbf{i}_{abcs} + \boldsymbol{\lambda}'_m \tag{3.50}$$

In terms of $qd0$ variables

$$(\mathbf{K}^r_s)^{-1}\boldsymbol{\lambda}^r_{qd0s} = \mathbf{L}_s(\mathbf{K}^r_s)^{-1}\mathbf{i}^r_{qd0s} + \boldsymbol{\lambda}'^r_m \tag{3.51}$$

Pre-multiplying by $\mathbf{K}^r_s$ yields

$$\begin{bmatrix} \lambda^r_{qs} \\ \lambda^r_{ds} \\ \lambda_{0s} \end{bmatrix} = \begin{bmatrix} L_{ls} + \frac{3}{2}L_{ms} & 0 & 0 \\ 0 & L_{ls} + \frac{3}{2}L_{ms} & 0 \\ 0 & 0 & L_{ls} \end{bmatrix} \begin{bmatrix} i^r_{qs} \\ i^r_{ds} \\ i_{0s} \end{bmatrix} + \lambda'^r_m \begin{bmatrix} 0 \\ 1 \\ 0 \end{bmatrix} \tag{3.52}$$

Since we will consider only balanced conditions, the $0s$ variables do not exist. We see from (3.40) that

$$\lambda^r_{qs} = \left(L_{ls} + \frac{3}{2}L_{ms}\right) i^r_{qs} \tag{3.53}$$

$$\lambda^r_{ds} = \left(L_{ls} + \frac{3}{2}L_{ms}\right) i^r_{ds} + \lambda'^r_m \tag{3.54}$$

and we must multiply L_{ms} by $\frac{3}{2}$ that is

$$L_{ss} = L_{ls} + \frac{3}{2}L_{ms} \tag{3.55}$$

3.3.4 Steady-state Analysis

The converter of the brushless dc drive is controlled so that the applied stator voltages are of the frequency of the electrical angular frequency of the rotor speed;

in other words $\omega_e = \omega_r$. Therefore, in the steady state, the ${}^{r}_{qs}$- and ${}^{r}_{ds}$ variables are constant and (3.37) and (3.38) become

$$V^r_{qs} = r_s I^r_{qs} + \omega_r L_{ss} I^r_{ds} + \omega_r \lambda'^r_m \tag{3.56}$$

$$V^r_{ds} = r_s I^r_{ds} - \omega_r L_{ss} I^r_{qs} \tag{3.57}$$

If we substitute (3.56) and (3.57) into (2.100) only if $\omega_r = \omega_e$, we have

$$\begin{aligned} \sqrt{2}\tilde{V}_{as} &= V^r_{qs} - jV^r_{ds} \\ &= r_s I^r_{qs} + \omega_r L_{ss} I^r_{ds} + \omega_r \lambda'^r_m - j(r_s I^r_{ds} - \omega_r L_{ss} I^r_{qs}) \end{aligned} \tag{3.58}$$

Now,

$$j\sqrt{2}\tilde{F}_{as} = F^r_{ds} + jF^r_{qs} \tag{3.59}$$

hence, (3.58) may be written as

$$\tilde{V}_{as} = (r_s + j\omega_r L_{ss})\tilde{I}_{as} + \frac{1}{\sqrt{2}}\omega_r \lambda'^r_m e^{j0} \tag{3.60}$$

We can define

$$\tilde{E}_a = \frac{1}{\sqrt{2}}\omega_r \lambda'^r_m e^{j0} \tag{3.61}$$

and write (3.60) as

$$\tilde{V}_{as} = (r_s + j\omega_r L_{ss})\tilde{I}_{as} + \tilde{E}_a \tag{3.62}$$

From (3.41), for a two-phase machine

$$\begin{aligned} T_e &= \frac{P}{2}\lambda'^r_m I^r_{qs} \\ &= \frac{P}{2}\lambda'^r_m \sqrt{2} I_s \cos\left[\theta_{esi}(0) - \theta_r(0)\right] \end{aligned} \tag{3.63}$$

It is important to point out that the load torque determines the rotor speed ω_r whereupon the converter makes $\omega_e = \omega_r$, which is the frequency of the applied stator voltages. Therefore, ω_r is used in (3.61) and (3.62) to call attention to this fact. For three-phase $L_{ms} = \frac{3}{2}L_{ms}$ and T_e is multiplied by $\frac{3}{2}$.

Example 3C ***Steady-state operation of a two-phase permanent-magnet ac machine***

Consider a four-pole two-phase permanent-magnet ac machine. The rotor angular velocity $\omega_{rm} = 50$ rad/s. $V_s = 40$ V, $I_s = 1$ A, $r_s = 4\ \Omega$, $L_{ss} = 20$ mH, and ϕ_v is equal to zero; thus, V^r_{ds} is controlled equal to zero. (a) Calculate $\tilde{E}_a$ and draw the phasor diagram, including the location of the stator and rotor poles, and (b) plot the magnetic fields as viewed running counterclockwise at ω_e which is ω_r.

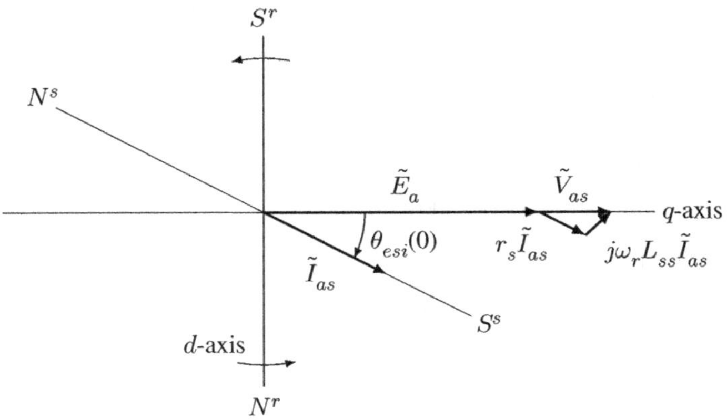

Figure 3C.1 Phasor diagram for steady-state operation of two-phase permanent-magnet ac machine.

Since V_{ds}^r is zero, $\tilde{V}_{as}$ (3.60) and $\tilde{E}_a$ (3.61) are in phase. Now, since $\tilde{V}_{as}$ is controlled to be in phase with $\tilde{E}_a$, (we'll see how this is done by inverter control in Chapter 4), the current lags $\tilde{V}_{as}$ (q axis) by the impedance phase angle. That is

$$\theta_{esi}(0) = -\tan^{-1}\frac{\left(\frac{4}{2}\right)(50)(20\times10^{-3})}{4} = -26.6^\circ \tag{3C.1}$$

From (3.60) and (3.61),

$$\begin{aligned}\tilde{E}_a &= \tilde{V}_{as} - (r_s + j\omega_r L_{ss})\tilde{I}_{as}\\ &= 40\underline{/0^\circ} - \left[4 + j\left(\frac{4}{2}\right)(50)(20\times10^{-3})\right] 1\underline{/-26.6^\circ}\\ &= 40 - (4 + j2)\underline{/-26.6^\circ} = 35.5\underline{/0^\circ}\text{ V}\end{aligned} \tag{3C.2}$$

The phasor diagram is given in Figure 3C.1; the stator and rotor mmfs are shown in Figure 3C.2.

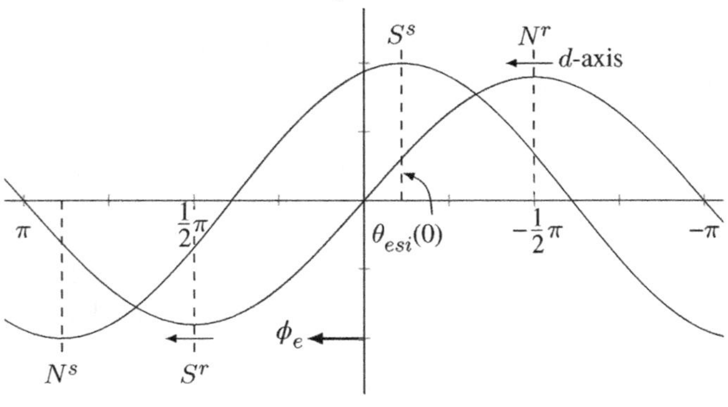

Figure 3C.2 Observing the stator and rotor rotating magnetic fields while running at $\omega_r(\omega_e)$ counterclockwise with 360° vision.

We see from Figures 3C.1 and 3C.2 that the rotor poles are being "pulled" by the stator poles and the direction of the torque is counterclockwise; motor action.

Example 3D ***Steady-state operation of a three-phase permanent-magnet machine***

A three-phase four-pole fractional-horsepower permanent-magnet ac machine has the following parameters $r_s = 3.4\ \Omega$, $L_{ls} = 1.1$ mH, $L_{ms} = 11$ mH, and $\lambda'_m = 0.0826$ V s/rad where $\lambda'_m = \lambda'^r_m$. The machine is wye-connected with balanced three-phase voltages. The *as*-phase voltage is $V_{as} = \sqrt{2}\,11.25\cos\omega_e t$. The actual steady-state rotor speed is 750 r/min. Calculate ω_e and T_e, and $\tilde{I}_{as}$ if $\phi_v = 0$ and draw a phasor diagram and show the stator and rotor poles.

The electrical angular velocity of the rotor if from (2.109)

$$\begin{aligned}\omega_r &= \frac{P}{2}\omega_{rm}\\ &= \frac{P}{2}\frac{(\text{r/min})(\text{rad/r})}{\text{s/min}}\\ &= \frac{4}{2}\frac{(750)(2\pi)}{60} = 50\pi \text{ rad/s}\end{aligned} \tag{3D.1}$$

We are firing the inverter so that the frequency of the voltages applied to the stator windings is equal to the electrical angular velocity of the rotor; therefore, $\omega_e = 50\pi$ rad/s. The voltage equations are identical for the two- and three-phase machines except for the $\frac{3}{2}L_{ms}$ in L_{ss} and $\frac{3}{2}$ times the torque. We need $\tilde{E}_a$, thus from (3.61)

$$\begin{aligned}\tilde{E}_a &= \frac{1}{\sqrt{2}}\omega_r\lambda'^r_m\underline{/0^\circ}\\ &= \frac{1}{\sqrt{2}}(50\pi)(0.0826)\underline{/0^\circ} = 9.18\underline{/0^\circ}\text{ V}\end{aligned} \tag{3D.2}$$

Now from (3.60)

$$\tilde{V}_{as} = (r_s + j\omega_e L_{ss})\tilde{I}_{as} + \tilde{E}_a$$

$$11.25\underline{/0^\circ} = \left(3.4 + j50\pi\left(1.1 + \frac{3}{2}11\right)\times 10^{-3}\right)\tilde{I}_{as} + 9.18\underline{/0^\circ}$$

$$\tilde{I}_{as} = \frac{(11.25 - 9.18)\underline{/0^\circ}}{3.4 + j2.76} = \frac{2.07\underline{/0^\circ}}{4.38\underline{/39.1^\circ}} = 0.473\underline{/-39.1^\circ}\text{ A} \tag{3D.3}$$

From (3.63)

$$\begin{aligned}T_e &= \left(\frac{3}{2}\right)\left(\frac{P}{2}\right)\lambda'^r_m\sqrt{2}I_s\cos[\theta_{esi}(0) - \theta_r(0)]\\ &= \left(\frac{3}{2}\right)\left(\frac{4}{2}\right)0.0826\sqrt{2}(0.473)\cos(-39.1^\circ - 0)\\ &= 0.129\text{ N m}\end{aligned} \tag{3D.4}$$

The phasor diagram is of the same form as Figure 3C.1.

SP3.3.1 The stator windings of the permanent-magnet ac machine shown in Figure 3.6 are open-circuited. The rotor is driven clockwise and $V_{as} = -10\sin 100t$. Determine V_{bs}. [$V_{bs} = -10\cos 100t$]

SP3.3.2 Determine λ'_m for SP3.3.1. [$\lambda'_m = 0.1$ V s/rad]

SP3.3.3 In the drive application of the permanent-magnet ac machine, the frequency of v_{as} and v_{bs} corresponds to the electrical angular velocity of the rotor. Therefore, assume that $I_{as} = \sqrt{2}I_s \cos[\omega_r t + \theta_{esi}(0)]$ and $I_{bs} = \sqrt{2}I_s \sin[\omega_r t + \theta_{esi}(0)]$ and $\theta_r = \omega_r t + \theta_r(0)$. Express T_e using (3.40). $\left\{\frac{P}{2}\lambda'_m \sqrt{2}I_s \cos[\theta_{esi}(0) - \theta_r(0)]\right\}$

SP3.3.4 Let $f_{as} = -\cos\theta_r$ and $f_{bs} = -\sin\theta_r$. Determine f^r_{qs} and f^r_{ds}. [$f^r_{qs} = -1; f^r_{ds} = 0$]

SP3.3.5 Let $f_{as} = \cos\theta_{esf}$, $f_{bs} = \sin\theta_{esf}$, and $\theta_r = \omega_r t$. Express θ_{esf} for $f^r_{qs} = 0$ and $f^r_{ds} = 1$. [$\theta_{esf} = \omega_r t - \frac{1}{2}\pi$]

SP3.3.6 If the speed ω_r is constant and if $I_{as} = \cos\theta_r$ and $I_{bs} = \sin\theta_r$, what will be the waveform of I^r_{qs} and I^r_{ds} (steady-state currents)? [dc]

SP3.3.7 Express ϕ_v if $\theta_{esv} = \omega_r t + \theta_{esv}(0)$ and $\theta_r = \omega_r t$. [$\phi_v = \theta_{esv}(0)$]

SP3.3.8 Start with (3.42) and obtain (3.41).

SP3.3.9 A four-pole three-phase permanent-magnet ac motor is operating with $\theta_r(0) = 0$, $V_{as} = 10\cos\omega_r t$, and $I_{as} = \cos(\omega_r t - 20°)$. Calculate λ'^r_m if $r_s = 4\ \Omega$, $L_{ss} = 0.01$ H, and $\omega_r = 146$ rad/s. [$\lambda'^r_m = 0.0393$ V s/rad]

SP3.3.10 Repeat 3.9 neglecting $\omega_r^2 L_{ss}^2$. [$\lambda'^r_m = 0.04427$ V s/rad]

SP3.3.11 Calculate (a) the input power P_e, (b) T_e, and (c) efficiency for the operating conditions given in SP3.3.9. [(a) $P_e = 14.1$ W; (b) $T_e = 0.111$ N m; (c) eff = 57.4%]

SP3.3.12 The expression for torque of a two-phase two-pole permanent-magnet ac machine is $T_e = \lambda'^r_m i^r_{qs}$. (a) What is the torque expression for a two-phase multipole permanent-magnet ac machine? (b) Multipole three-phase machine? [(a) $T_e = \frac{P}{2}\lambda'^r_m i^r_{qs}$, (b) $\frac{3}{2}$ times (a)]

SP3.3.13 Why is $L_{ss} = L_{ls} + \frac{3}{2}L_{ms}$ for a three-phase machine. [Mutual coupling between phases and $i_{as} + i_{bs} + i_{cs} = 0$]

3.4 Symmetrical Induction Machines

Although the induction machine is generally used as a means to convert electric power to mechanical work, it can operate as either a motor or a generator. Three-phase induction motors are commonly used in large-horsepower applications, for example, pump drives, steel mill drives, hoist drives, vehicle drives, and as a generator in wind turbine and low-head hydro applications. Single-phase induction motors, which develop torque in a manner similar to the multiphase induction motor, are used in many household appliances. The analysis of symmetrical two- and three-phase induction machines is essentially the same. Therefore, we will focus our attention on the two-phase machine, since this enables us to become familiar with the theory and performance of induction machines without becoming inundated with three-phase trigonometric manipulations.

A cutaway of a four-pole three-phase 7.5-hp 460-V squirrel-cage induction motor is shown in Figure 3.9. It is an enclosed fan-cooled severe-duty motor for use in the chemical, paper, cement, and mining industries. A disassembled four-pole

Figure 3.9 Four-pole three-phase 6.5-Hp 460-V severe-duty, squirrel-cage induction motor. Source: Courtesy of General Electric.

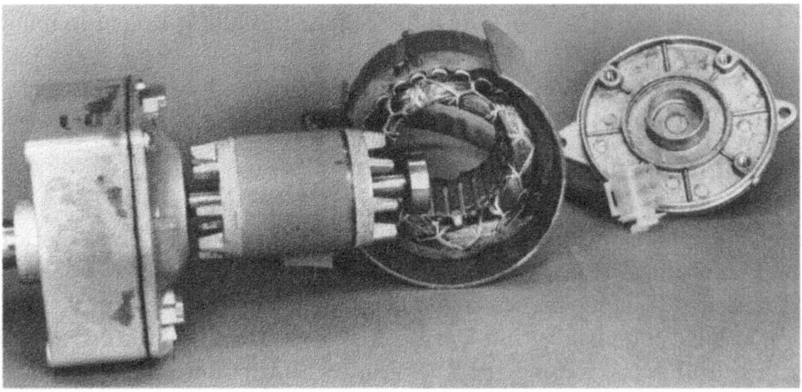

Figure 3.10 Four-pole two-phase 1/10-Hp 115-V induction motor with reduction gear.

two-phase $\frac{1}{10}$-hp 115-V induction motor, which is used in low-power control applications, is shown in Figure 3.10. Also shown in Figure 3.10 is the case that houses the speed-reduction gears. The rotors of the induction motors shown in Figures 3.9 and 3.10 are referred to as "squirrel-cage" rotors. Although it is difficult to discern, the squirrel cage rotors are made up of laminated punched steel with aluminum bars die casted in the openings of the laminated rotor and the bars terminated at each end of the rotor in an aluminum ring, which is visible in Figures 3.9 and 3.10. The protrusions from the aluminum rings are for cooling purposes. If we remove the steel laminations, the remaining aluminum bars and end rings resemble the rotor (blades) of a "squirrel-cage fan." The question often arises as to the validity of considering these short-circuited bars producing a rotating magnetic field. It would require considerable work to show this analytically; however, let us try to talk our way through it. Recall that the rotating magnetic field produced by the stator comes about due to the sinusoidally distributed windings and the balanced stator currents. The currents induced in the short-circuited rotor bars are unrestricted, and one can argue that the stator mmf will induce rotor currents such that the resulting rotor rotating magnetic field (mmf_r) would be sinusoidal in waveform as the inducing mmf (mmf_s) with an angular velocity equal to the difference in rotor speed from synchronous speed. This is essentially what would occur if the rotor were equipped with symmetrical sinusoidally distributed windings with currents of an angular frequency of $\omega_e - \omega_r$. Clearly, this is a first-order explanation or approximation, but it is commonly accepted, so we'll follow suit.

3.4.1 Two-phase Induction Machine

A two-pole two-phase induction machine is shown in Figure 3.11. It is assumed that the stator windings may be portrayed by orthogonal, sinusoidally distributed

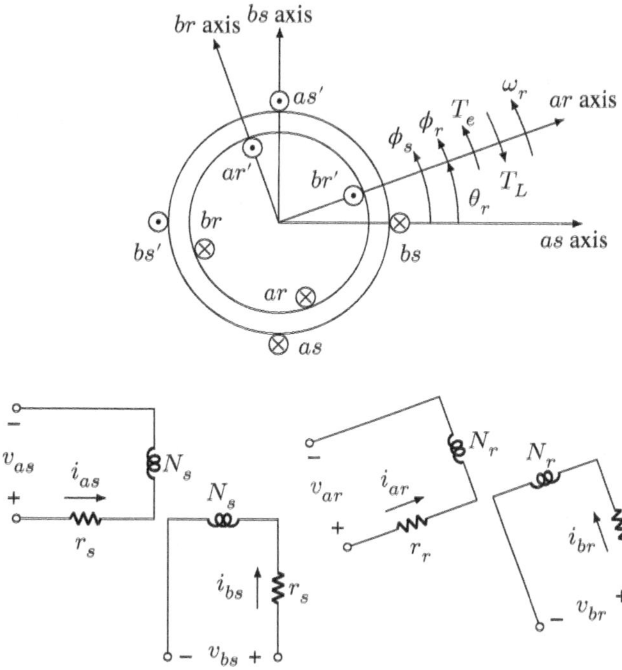

Figure 3.11 A two-pole, two-phase, symmetrical machine.

windings as described in Chapter 2. It is convenient to assume that the rotor of the two-pole two-phase induction machine may also be portrayed electrically by two sinusoidally distributed windings displaced 90°. Hence, for our present purposes, we will consider that the *ar* and *br* windings are sinusoidally distributed, each with the same number of turns and the same winding resistance. Thus, both the stator and rotor are symmetrical. For this reason, this device is often referred to as a *symmetrical machine.*

Note that the air gap distance between the stator and rotor is uniform. The rotor windings of an induction machine are generally short-circuited ($v_{ar} = v_{br} = 0$) with only the stator windings connected to a source. In this case, the machine is said to be single-fed. In some special applications, such as wind turbines, both the rotor and stator windings are connected to sources. In particular, the rotor windings are connected to a stationary multi-phase source by a brush and slip-ring arrangement. In this case, the machine is double-fed. We will not deal with a double-fed machine.

As established in Chapter 2, the angular displacement about the stator is denoted ϕ_s, and it is referenced to the *as* axis. We see from Figure 3.11 that the angular displacement about the rotor is denoted ϕ_r and it is referenced to the *ar* axis. The angular velocity of the rotor is ω_r and θ_r is its angular position between the *ar* and *as* axes. The electromechanical torque T_e and the load torque T_L are also indicated in Figure 3.11. The torque T_e is assumed to be positive in the direction of increasing θ_r, whereas the load torque T_L is positive in the opposite direction (opposing rotation).

Let us take a few minutes to review the things we have already learned about an induction machine. In our analysis, we will assume that the stator and equivalent windings of the rotor are sinusoidally distributed and symmetrical. Since the rotor currents of frequency $\omega_e - \omega_r$ are flowing in circuits that are rotating at ω_r, Tesla's rotating magnetic field of the stator (mmf_s) and the rotor (mmf_r) rotate about the air gap at ω_e. Therefore, mmf_s and mmf_r are always stationary relative to each other in the steady state. A constant torque is produced due to the force between these magnetic systems.

Since the currents induced in the rotor circuits are due to a difference in the speed of the rotor and the stator rotating magnetic field, rotor currents would not be induced when the angular velocity of the rotor is equal to the angular velocity of the stator rotating magnetic field ($\omega_r = \omega_e$). Thus, the symmetrical machine will not develop a torque at synchronous speed; only when $\omega_r \neq \omega_e$ will rotor currents be induced and a torque produced. This, of course, is completely opposite to a synchronous machine which produces an average torque only when $\omega_r = \omega_e$. By now you have probably guessed that the induction machine is so named from the fact that torque is produced as a result the rotor currents being "induced" by the stator rotating magnetic field. It is interesting that the induction machine was first referred to as a rotating transformer. This is understandable, but we'll see this more clearly when we derive the steady-state equivalent circuit for the induction machine and find that it is similar to that of a transformer. Also, we should mention that generator action occurs if the stator windings are connected to a source, thus establishing a rotating magnetic field, and the rotor, with its short circuited windings, is driven above synchronous speed.

In summary, the induction or symmetrical machine has two sets of symmetrical circuits, one on the stator and the other on the rotor. The set on the rotor is short circuited and the frequency of the balanced set of rotor currents is $\omega_e - \omega_r$. This results in the mmf_s and the mmf_r that rotate in unison, and torque is produced due to the interaction of these two mmfs.

3.4.2 Symmetrical Rotor Windings

The derivation of the air-gap mmf of symmetrical rotating circuits parallels that of the stationary circuits. As shown in Figure 3.12, the *ar* and *br* windings are orthogonal and if they are identical and sinusoidally distributed, Tesla's rotating

Figure 3.12 Two-phase rotating, identical, sinusoidally distributed symmetrical windings.

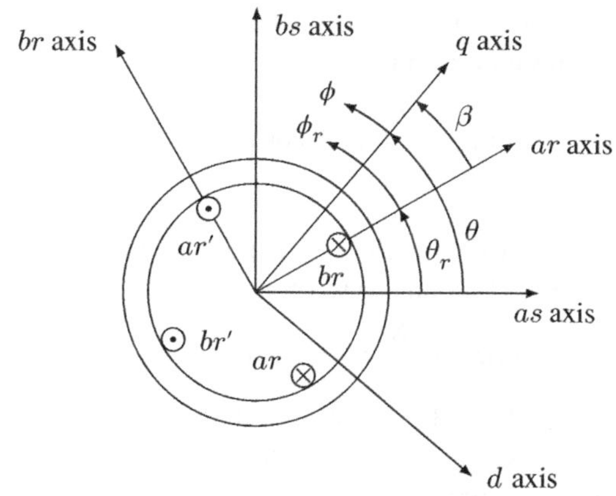

magnetic field of these symmetrical windings may be expressed.

$$\text{mmf}_r = \frac{N_r}{2}(i_{ar}\cos\phi_r + i_{br}\sin\phi_r) \tag{3.64}$$

where N_r is the equivalent number of turns of the rotor windings and ϕ_r is its angular position from the *ar* axis, which is positioned by θ_r from the *as* axis. In Figure 3.12, the positions θ_r and θ are referenced to the *as* axis.

Now, the positions are related as

$$\phi_r = \beta + \phi \tag{3.65}$$

Substituting (3.65) into (3.64) yields

$$\text{mmf}_r = \frac{N_r}{2}[i_{ar}\cos(\beta + \phi) + i_{br}\sin(\beta + \phi)] \tag{3.66}$$

We will return to (3.66) in a moment; however, let us continue our focus on the rotating air-gap mmf during balanced steady-state operation.

Now,

$$\beta = \theta - \theta_r \tag{3.67}$$

and during steady-state operation

$$\beta = (\omega - \omega_r)t + \theta(0) - \theta_r(0) \tag{3.68}$$

During balanced steady-state operation of an induction motor, the rotor windings are short-circuited and the rotor currents, which are induced by mmf_s, are balanced with the angular frequency of $(\omega_e - \omega_r)$. Thus,

$$I_{ar} = \sqrt{2}I_r\cos[(\omega_e - \omega_r)t + \theta_{eri}(0)] \tag{3.69}$$

$$I_{br} = \sqrt{2}I_r\sin[(\omega_e - \omega_r)t + \theta_{eri}(0)] \tag{3.70}$$

It is important to recall that I_{ar} and I_{br} are zero if $\omega_r = \omega_e$.

If we now substitute (3.67), (3.69), and (3.70) into (3.66), we will obtain an expression for mmf_r

$$\text{mmf}_r = \frac{N_r}{2}\sqrt{2}I_r \cos\left[(\omega_e - \omega_r - \omega + \omega_r)t + \theta_{eri}(0) - \theta(0) + \theta_r(0) - \phi\right] \tag{3.71}$$

Now, we will select $\theta(0)$ and $\theta_r(0)$ to be zero unless otherwise specified; whereupon, (3.71) may be written

$$\text{mmf}_r = \frac{N_r}{2}\sqrt{2}I_r \cos[(\omega_e - \omega)t + \theta_{eri}(0) - \phi] \tag{3.72}$$

Equation (3.72) is similar in form to mmf_s given by (2.45). This is a very important observation. If the rotor currents are balanced and if the angular frequency of these currents is $(\omega_e - \omega_r)$, then the stator mmf_s and the rotor mmf_r will travel around the air gap at the same angular velocity regardless of the actual speed of the rotor if $\omega_r \neq \omega_e$. If, for example, we observe mmf_s and mmf_r as a stationary observer ($\omega = 0$), both mmf_s^s and mmf_r^s would be traveling counterclockwise at ω_e relative to us if $\omega_r \neq \omega_e$. If we are riding on the rotor ($\omega = \omega_r$), mmf_s^r and mmf_r^r would be traveling at $(\omega_e - \omega_r)$ counterclockwise relative to us. If we are running at ω_e, mmf_s^e and mmf_r^e would appear to us as constant space sinusoids of ϕ_e during the steady state. Since mmf_s and mmf_r are traveling in unison, a constant torque (power) is produced. We are starting to uncover the basis of Tesla's contribution to electric power generation and utilization.

A constant torque can be produced only if mmf_s and mmf_r are rotating at the same angular velocity and $\omega_r \neq \omega_e$. This is the case for balanced steady-state operation of a symmetrical machine. The magnitude of the constant torque is determined by the number of turns of stator or rotor windings, the magnitude of the stator and rotor currents, and the relative position of mmf_s and mmf_r. The relative position of the phase between the mmfs, $\theta_{esi}(0)$ and $\theta_{eri}(0)$, is easily discerned in the synchronous reference frame, i.e. mmf_s^e and mmf_r^e

$$\text{mmf}_s^e = \frac{N_s}{2}\sqrt{2}I_s \cos[\theta_{esi}(0) - \phi_e] \tag{3.73}$$

$$\text{mmf}_r^e = \frac{N_r}{2}\sqrt{2}I_r \cos[\theta_{eri}(0) - \phi_e] \tag{3.74}$$

In the dc machine, torque is produced by the interaction of two stationary, orthogonal mmfs. A synchronous machine develops a constant torque only at synchronous speed, $\omega_r = \omega_e$. This occurs due to the fact that the rotor field, which is stationary with respect to the rotor, is rotating in unison with Tesla's rotating magnetic field. This, of course, is necessary to produce an average torque. Now, the induction or symmetrical machine produces torque at any rotor speed except when $\omega_r = \omega_e$ and current is not induced in the short-circuited rotor circuits. In this case, mmf_r disappears and mmf_s has no other mmf with which to interact.

3.4.3 Substitute Variables for Symmetrical Rotating Circuits

The reference frame derivation may be deemphasized and in-depth study resumed with the Section 3.4.5. The voltage equations expressed for the stator circuits in the arbitrary reference frame are given by (2.86) and (2.87). We need similar equations for the symmetrical rotor circuits. The transformation for the rotor variables to the arbitrary reference frame follows that set forth in Section 2.7 for the stationary circuits. In particular, if (3.66) is substituted into (3.64) and after a few trigonometric manipulations, we have

$$\text{mmf}_r = \frac{N_r}{2}\cos\phi(i_{ar}\cos\beta + i_{br}\sin\beta) + \frac{N_r}{2}\sin\phi(i_{ar}\cos\beta + i_{br}\sin\beta) \quad (3.75)$$

If we let $\phi = 0$, we obtain i_{qr} and $\phi = -\frac{\pi}{2}$ we obtain i_{dr}. Equation (3.75) may now be written as

$$\text{mmf}_r = \frac{N_r}{2}(i_{qr}\cos\phi - i_{dr}\sin\phi) \quad (3.76)$$

Figure 3.13 depicts the fictitious *qr* and *dr* windings for the rotor. It is important to mention that if a device has symmetrical stator windings and symmetrical rotor windings, as is the case of an induction or symmetrical machine, the stator and rotor windings have symmetrical fictitious windings in the arbitrary reference frame. Moreover, in the arbitrary reference frame, the fictitious windings are all stationary relative to each other. This is significant since we will find that stator and rotor position-dependent mutual inductances are eliminated in all reference frames in the case of the symmetrical machine.

The transformation for the rotating symmetrical circuits becomes

$$\begin{bmatrix} f_{qr} \\ f_{dr} \end{bmatrix} = \begin{bmatrix} \cos\beta & \sin\beta \\ \sin\beta & -\cos\beta \end{bmatrix} \begin{bmatrix} f_{ar} \\ f_{br} \end{bmatrix} \quad (3.77)$$

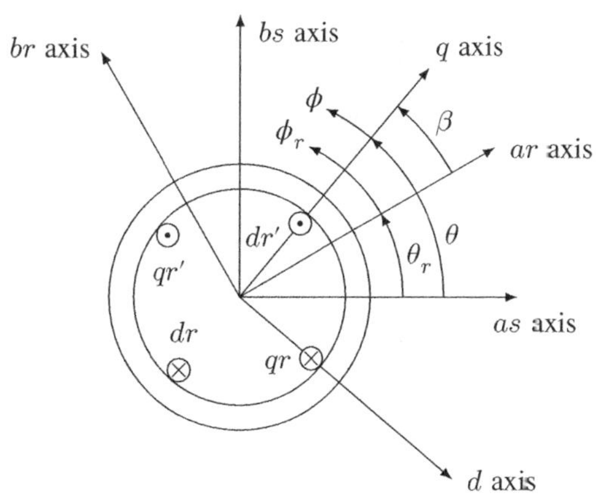

Figure 3.13 The fictitious *qr* and *dr* windings.

or

$$\mathbf{f}_{qdr} = \mathbf{K}_r \mathbf{f}_{abr} \tag{3.78}$$

and

$$(\mathbf{K}_r)^{-1} = \mathbf{K}_r \tag{3.79}$$

The substitute arbitrary reference frame voltage equations for a resistive–inductive circuit become

$$v_{qr} = r_r i_{qr} + (\omega - \omega_r)\lambda_{dr} + p\lambda_{qr} \tag{3.80}$$

$$v_{dr} = r_r i_{dr} - (\omega - \omega_r)\lambda_{qr} + p\lambda_{dr} \tag{3.81}$$

These substitute voltage equations are very similar in form to those for stationary circuits given by (2.86) and (2.87), in particular, $\omega - \omega_r$ replaces ω.

The extension to three-phase rotating circuits is straightforward and will not be given; however, before proceeding, it is interesting to note that if θ_r and ω_r are zero, the transformation given by (3.77) becomes the transformation to the arbitrary reference frame for the stationary circuits. Therefore, (3.77) is actually the only transformation that is needed for stationary or rotating symmetrical circuits.

We have two things to do before we can come up with equivalent circuits. In particular, we must refer the rotor variables to the stator windings and next we must express $\boldsymbol{\lambda}_{qds}$ and $\boldsymbol{\lambda}'_{qdr}$. We are not going to go through all of the algebra and matrix manipulations necessary to accomplish this; instead, we are going to talk our way to the equivalent circuits. First, the turn ratios necessary to refer the rotor variables to the stator are

$$\mathbf{i}'_{abr} = \frac{N_r}{N_s}\mathbf{i}_{abr} \tag{3.82}$$

$$\mathbf{v}'_{abr} = \frac{N_s}{N_r}\mathbf{v}_{abr} \tag{3.83}$$

$$\boldsymbol{\lambda}'_{abr} = \frac{N_s}{N_r}\boldsymbol{\lambda}_{abr} \tag{3.84}$$

The substitute variables may now be written in the arbitrary reference frame as

$$v_{qs} = r_s i_{qs} + \omega\lambda_{ds} + p\lambda_{qs} \tag{3.85}$$

$$v_{ds} = r_s i_{ds} - \omega\lambda_{qs} + p\lambda_{ds} \tag{3.86}$$

$$v'_{qr} = r'_r i'_{qr} + (\omega - \omega_r)\lambda'_{dr} + p\lambda'_{qr} \tag{3.87}$$

$$v'_{dr} = r'_r i'_{dr} - (\omega - \omega_r)\lambda'_{qr} + p\lambda'_{dr} \tag{3.88}$$

where

$$r'_r = \left(\frac{N_s}{N_r}\right)^2 r_r \tag{3.89}$$

Equations (3.85) and (3.86) are (2.86) and (2.87); (3.87) and (3.88) are referred versions of (3.80) and (3.81).

Although (3.85)–(3.88) are valid for a nonlinear magnetic system, we generally assume that the system is magnetically linear, whereupon the flux linkages may be expressed as linear functions of inductances and currents. In particular, we can write

$$\lambda_{as} = L_{asas}i_{as} + L_{asbs}i_{bs} + L_{asar}i_{ar} + L_{asbr}i_{br} \tag{3.90}$$

$$\lambda_{bs} = L_{bsas}i_{as} + L_{bsbs}i_{bs} + L_{bsar}i_{ar} + L_{bsbr}i_{br} \tag{3.91}$$

$$\lambda_{ar} = L_{aras}i_{as} + L_{arbs}i_{bs} + L_{arar}i_{ar} + L_{arbr}i_{br} \tag{3.92}$$

$$\lambda_{br} = L_{bras}i_{as} + L_{brbs}i_{bs} + L_{brar}i_{ar} + L_{brbr}i_{br} \tag{3.93}$$

which may be written as

$$\begin{bmatrix} \mathbf{I}_{abs} \\ \mathbf{I}_{abr} \end{bmatrix} = \begin{bmatrix} \mathbf{L}_s & \mathbf{L}_{sr} \\ (\mathbf{L}_{sr})^T & \mathbf{L}_r \end{bmatrix} \begin{bmatrix} \mathbf{i}_{abs} \\ \mathbf{i}_{abr} \end{bmatrix} \tag{3.94}$$

where

$$\mathbf{L}_s = \begin{bmatrix} L_{ss} & 0 \\ 0 & L_{ss} \end{bmatrix} \tag{3.95}$$

$$\mathbf{L}_r = \begin{bmatrix} L_{rr} & 0 \\ 0 & L_{rr} \end{bmatrix} \tag{3.96}$$

$$\mathbf{L}_{sr} = L_{sr} \begin{bmatrix} \cos\theta_r & -\sin\theta_r \\ \sin\theta_r & \cos\theta_r \end{bmatrix} \tag{3.97}$$

and

$$L_{ss} = L_{ls} + L_{ms} \tag{3.98}$$

$$L_{rr} = L_{lr} + L_{mr} \tag{3.99}$$

Now all inductances except the leakage inductances have a common term $\mathcal{R}_m$ and can be related using turn ratios. Therefore, we can relate all inductances to the stator magnetizing inductance which means that all currents are flowing in L_{ms}. The work up to Figure 3.14 is a manipulation of the inductances and turns ratios and may be deemphasized.

Referring (3.94) to the stator winding (N_s) yields

$$\begin{bmatrix} \mathbf{i}_{abs} \\ \mathbf{i}'_{abr} \end{bmatrix} = \begin{bmatrix} \mathbf{L}_s & \frac{N_s}{N_r}\mathbf{L}_{sr} \\ \frac{N_s}{N_r}(\mathbf{L}_{sr})^T & \mathbf{L}'_r \end{bmatrix} \begin{bmatrix} \mathbf{i}_{abs} \\ \mathbf{i}'_{abr} \end{bmatrix} \tag{3.100}$$

where

$$\mathbf{L}'_r = \left(\frac{N_s}{N_r}\right)^2 \mathbf{L}_r = \begin{bmatrix} L'_{rr} & 0 \\ 0 & L'_{rr} \end{bmatrix} \tag{3.101}$$

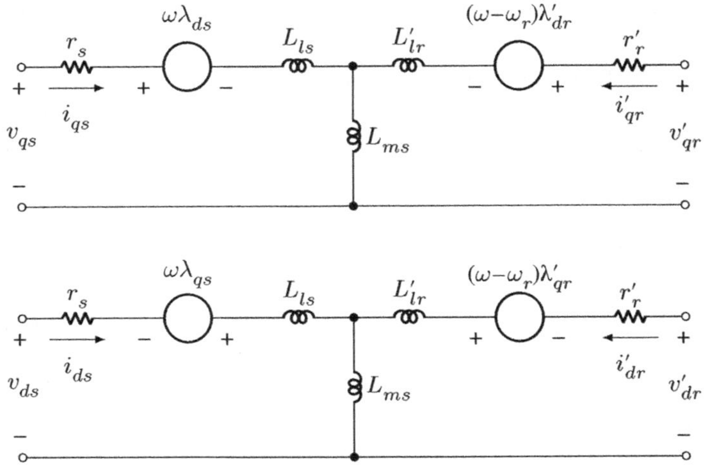

Figure 3.14 Arbitrary reference frame equivalent circuits for a two-phase, symmetrical induction machine.

Since L_{ms}, L_{mr}, and L_{sr} all have the common term R_m and $L'_{lr} = \left(\frac{N_s}{N_r}\right)^2 L_{lr}$,

$$\begin{aligned} L'_{rr} &= L'_{lr} + \left(\frac{N_s}{N_r}\right)^2 L_{mr} \\ &= L'_{lr} + L_{ms} \end{aligned} \tag{3.102}$$

Note that

$$\frac{N_s}{N_r}\mathbf{L}_{sr} = \frac{N_s}{N_r}L_{sr}\begin{bmatrix} \cos\theta_r & -\sin\theta_r \\ \sin\theta_r & \cos\theta_r \end{bmatrix} \tag{3.103}$$

and

$$\frac{N_s}{N_r}L_{sr} = L_{ms} \tag{3.104}$$

Hence, (3.103) may be expressed in terms of L_{ms} and, for compactness, we will define L'_{sr} as

$$\begin{aligned} \mathbf{L}'_{sr} &= \frac{N_s}{N_r}\mathbf{L}_{sr} \\ &= L_{ms}\begin{bmatrix} \cos\theta_r & -\sin\theta_r \\ \sin\theta_r & \cos\theta_r \end{bmatrix} \end{aligned} \tag{3.105}$$

Thus, (3.100) becomes

$$\begin{bmatrix} \mathbf{I}_{abs} \\ \mathbf{I}'_{abr} \end{bmatrix} = \begin{bmatrix} \mathbf{L}_s & \mathbf{L}'_{sr} \\ (\mathbf{L}'_{sr})^T & \mathbf{L}_r \end{bmatrix}\begin{bmatrix} \mathbf{L}_{abs} \\ \mathbf{L}'_{abr} \end{bmatrix} \tag{3.106}$$

The flux linkage equations for a magnetically linear system given by (3.106) are in terms of *as*, *bs*, *ar*′, and *br*′ variables. In the arbitrary reference frame, (3.106) becomes

$$\begin{bmatrix} \mathbf{i}_{qds} \\ \mathbf{i}'_{qdr} \end{bmatrix} = \begin{bmatrix} \mathbf{K}_s\mathbf{L}_s(\mathbf{K}_s)^{-1} & \mathbf{K}_s\mathbf{L}'_{sr}(\mathbf{K}_r)^{-1} \\ \mathbf{K}_r(\mathbf{L}'_{sr})^T(\mathbf{K}_s)^{-1} & \mathbf{K}_r\mathbf{L}'_r(\mathbf{K}_r)^{-1} \end{bmatrix} \begin{bmatrix} \mathbf{i}_{qds} \\ \mathbf{i}'_{gdr} \end{bmatrix} \tag{3.107}$$

which becomes

$$\begin{bmatrix} \mathbf{i}_{qds} \\ \mathbf{i}'_{qdr} \end{bmatrix} = \begin{bmatrix} \mathbf{L}_s & \mathbf{L}_{ms} \\ \mathbf{L}_{ms} & \mathbf{L}'_r \end{bmatrix} \begin{bmatrix} \mathbf{i}_{qds} \\ \mathbf{i}'_{qdr} \end{bmatrix} \tag{3.108}$$

Since $L_{ss} = L_{ls} + L_{ms}$ and $L'_{rr} = L'_{lr} + L_{ms}$, (3.108) may be expressed as

$$\lambda_{qs} = L_{ls}i_{qs} + L_{ms}(i_{qs} + i'_{qr}) \tag{3.109}$$

$$\lambda_{ds} = L_{ls}i_{ds} + L_{ms}(i_{ds} + i'_{dr}) \tag{3.110}$$

$$\lambda'_{qr} = L'_{lr}i'_{qr} + L_{ms}(i_{qs} + i'_{qr}) \tag{3.111}$$

$$\lambda'_{dr} = L'_{lr}i'_{dr} + L_{ms}(i_{ds} + i'_{dr}) \tag{3.112}$$

Equations (3.85)–(3.89) and (3.109)–(3.112) suggest the equivalent circuits shown in Figure 3.14. It is interesting that for a symmetrical machine the rotor position-dependent terms of the mutual inductances are eliminated in all frames of reference. We would expect this since the stator windings are symmetrical and the rotor windings are also symmetrical and we are able to transform the variables to a common frame of reference; the arbitrary reference frame.

3.4.4 Torque

From (2.32) for a *P*-pole machine

$$T_e(\mathbf{i}, \theta_r) = \frac{P}{2}\frac{\partial W_c(\mathbf{i}, \theta_r)}{\partial \theta_r} \tag{3.113}$$

In a magnetically linear system, the energy in the coupling field W_f and the coenergy W_c are equal. The field energy can be expressed as

$$\begin{aligned} W_f(\mathbf{i}, \theta_r) = {} & \frac{1}{2}L_{ss}i_{as}^2 + \frac{1}{2}L_{ss}i_{bs}^2 + \frac{1}{2}L'_{rr}i'^2_{ar} + \frac{1}{2}L'_{rr}i'^2_{br} \\ & + L_{ms}i_{as}i'_{ar}\cos\theta_r - L_{ms}i_{as}i'_{br}\sin\theta_r \\ & + L_{ms}i_{bs}i'_{ar}\sin\theta_r + L_{ms}i_{bs}i'_{br}\cos\theta_r \end{aligned} \tag{3.114}$$

Since $W_f = W_c$, substituting (3.114) into (3.113) yields the electromagnetic torque for a magnetically linear *P*-pole two-phase symmetrical induction machine. In particular,

$$T_e = -\frac{P}{2}L_{ms}[(i_{as}i'_{ar} + i_{bs}i'_{br})\sin\theta_r + (i_{as}i'_{br} - i_{bs}i'_{ar})\cos\theta_r] \tag{3.115}$$

The torque and rotor speed are related by

$$T_e = J\frac{d\omega_{rm}}{dt} + B_m\omega_{rm} + T_L \tag{3.116}$$

where ω_{rm} is the actual rotor speed. For a P-pole machine, since $\omega_{rm} = (2/P)\omega_r$, thus

$$T_e = J\frac{2}{P}\frac{d\omega_r}{dt} + B_m\frac{2}{P}\omega_r + T_L \tag{3.117}$$

where J is the inertia of the rotor and, in some cases, the shaft-connected load. The first term on the right-hand side is the inertial torque. The units of J are kilogram meter^2 (kg m^2) or joules second^2 (J s^2). Often, the inertia is given as a quantity called WR^2 expressed in units of pound-mass feet^2 (lbm ft^2). The load torque T_L is positive for a torque load on the shaft of the induction machine, as shown in Figure 3.11. The constant B_m is a damping coefficient associated with the rotational system of the machine and mechanical load. It has the units of N m s/rad of mechanical rotation, and it is generally small and often neglected.

Equation (3.115) is an expression for the electromagnetic torque in machine currents i_{as}, i_{bs}, i'_{ar}, and i'_{br} for a magnetically linear system with the rotor windings short circuited. If we express the torque in terms of the arbitrary reference frame currents, the electromagnetic torque for a magnetically linear system may be expressed

$$T_e = \frac{P}{2}L_{ms}(i_{qs}i'_{dr} - i_{ds}i'_{qr}) \tag{3.118}$$

where T_e is positive for motor action. For a three-phase machine, the torque equation must be multiplied by $\frac{3}{2}$. Also, for a magnetically linear assumption, L_{ms} in the flux linkage equations must be multiplied by $\frac{3}{2}$.

The dynamic response of induction motor during free acceleration (no load) is shown in Figure 3.15. Figure 3.16 is the torque versus the rotor speed for free acceleration. Why isn't the torque on the torque versus rotor speed plot (Figure 3.16) sinusoidal in waveform?

The plots shown in Figure 3.17 are for the same free acceleration as viewed from the stationary reference frame where the rotor windings are fictitious stationary windings. There are several things to note; first, i^s_{ds} equals a negative i_{bs}; second, torque is the same regardless of the reference frame from which it is being viewed; and third, all voltages and currents are varying at the same frequency. This is necessary since the change of variables provides a portrayal of the rotating mmfs from the stationary reference frame, which is of ω_e frequency.

The free acceleration as viewed from the synchronous reference frame is shown in Figure 3.18. Here, both the stator and rotor windings become fictitious windings rotating at ω_e with constant steady-state currents. We see the torque being the same as in the other reference frames.

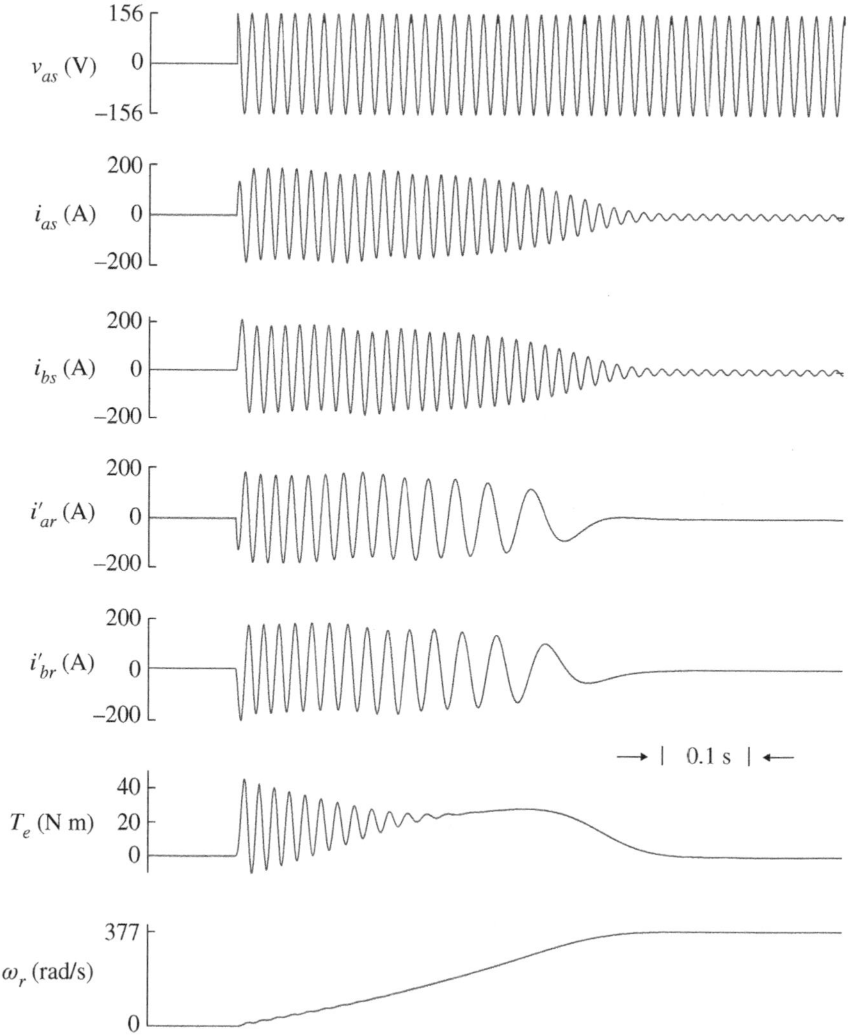

Figure 3.15 Free-acceleration characteristics of a two-pole two-phase 5-hp induction motor – machine variables.

Example 3E ***Write the voltage equations in the synchronously rotating reference frame***

From Figure 3.14 with $\omega = \omega_e$.

$$v_{qs}^e = r_s i_{qs}^e + \omega_e (L_{ss} i_{ds}^e + L_{ms} i'^e_{dr}) + L_{ss}\frac{di_{qs}^e}{dt} + L_{ms}\frac{di'^e_{qr}}{dt} \tag{3E.1}$$

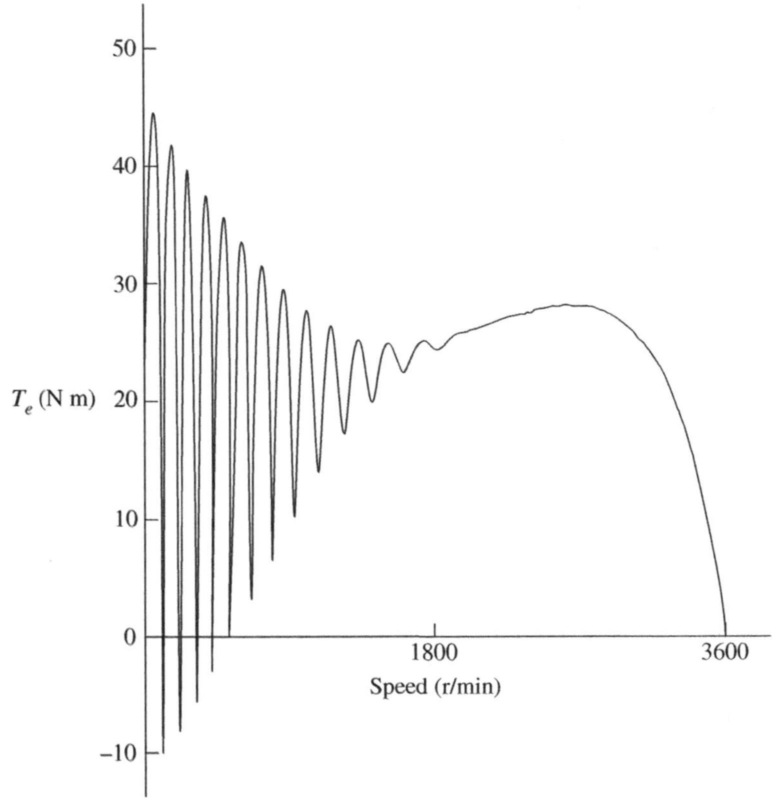

Figure 3.16 Torque versus speed during free acceleration shown in Figure 3.15.

$$v_{ds}^e = r_s i_{ds}^e - \omega_e (L_{ss} i_{qs}^e + L_{ms} i'^e_{qr}) + L_{ss} \frac{di_{ds}^e}{dt} + L_{ms} \frac{di'^e_{dr}}{dt} \tag{3E.2}$$

$$v'^e_{qr} = r'_r i'^e_{qr} + (\omega_e - \omega_r)(L'_{rr} i'^e_{dr} + L_{ms} i_{ds}^e) + L'_{rr} \frac{di'^e_{qr}}{dt} + L_{ms} \frac{di_{qs}^e}{dt} \tag{3E.3}$$

$$v'^e_{dr} = r'_r i'^e_{dr} - (\omega_e - \omega_r)(L'_{rr} i'^e_{qr} + L_{ms} i_{qs}^e) + L'_{rr} \frac{di'^e_{dr}}{dt} + L_{ms} \frac{di_{ds}^e}{dt} \tag{3E.4}$$

3.4.5 Phasors and Steady-state Equivalent Circuit

In this subsection, we will set forth the instantaneous and steady-state phasors and the steady-state equivalent circuit. The instantaneous phasor equations may be obtained by a procedure used in Chapter 2. In particular, we will substitute the voltage equations in the synchronously rotating reference frame given by

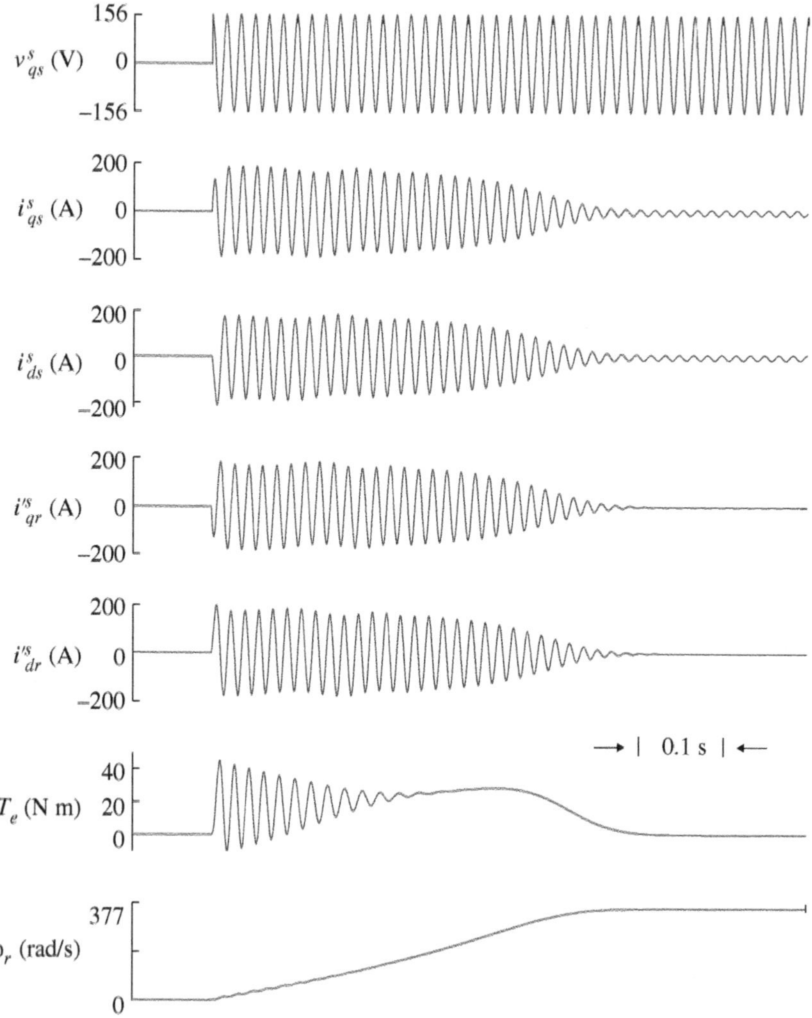

Figure 3.17 Same as Figure 3.15 – stationary reference variables.

(3E.1)–(3E.4) into

$$\tilde{f}_{as} = f_{qs}^e - jf_{ds}^e \tag{3.119}$$

$$\tilde{f}'_{ar} = f'^e_{qr} - jf'^e_{dr} \tag{3.120}$$

The instantaneous phasor voltage equations from Example 3E become,

$$\tilde{v}_{as} = r_s\tilde{i}_{as} + j\omega_e L_{ss}\tilde{i}_{as} + j\omega_e L_{ms}\tilde{i}'_{ar} + pL_{ss}\tilde{i}_{as} + pL_{ms}\tilde{i}'_{ar} \tag{3.121}$$

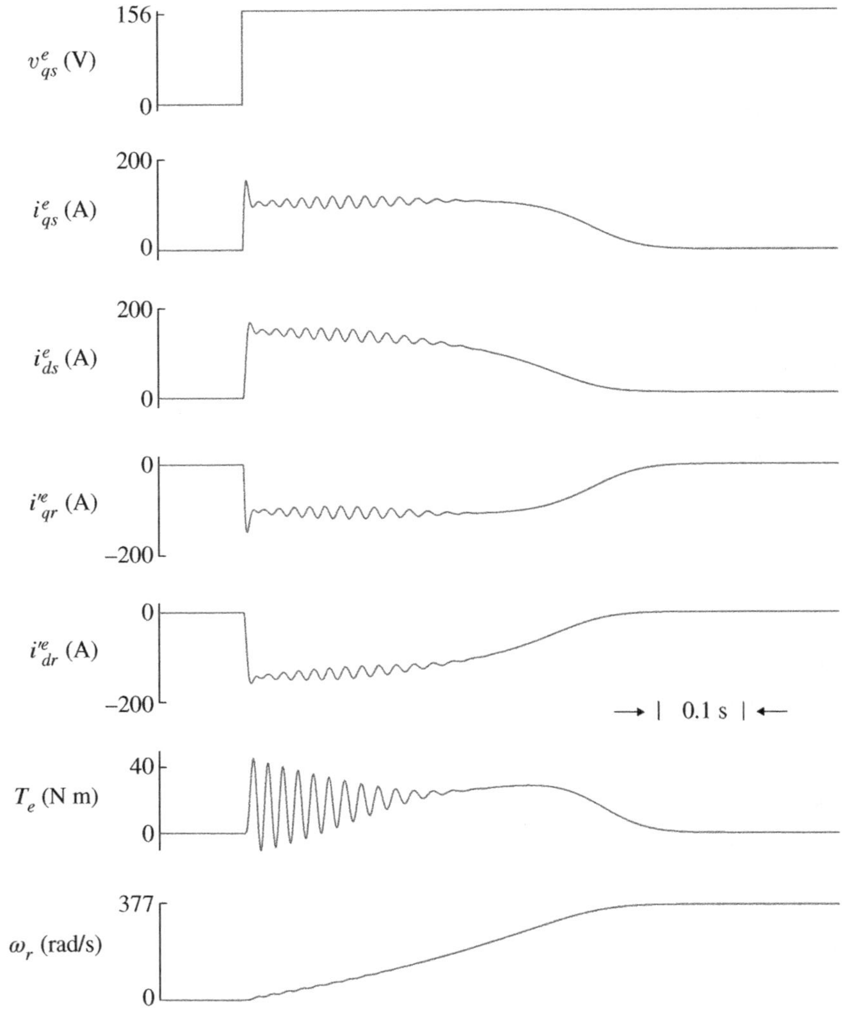

Figure 3.18 Same as Figure 3.15 – synchronously rotating reference frame variables.

$$\tilde{v}'_{ar} = r'_s\tilde{i}'_{ar} + j(\omega_e - \omega_r)L'_{rr}\tilde{i}'_{ar} + j(\omega_e - \omega_r)L_{ms}\tilde{i}_{as} + pL'_{rr}\tilde{i}'_{ar} + pL_{ms}\tilde{i}_{as} \tag{3.122}$$

For steady-state conditions the last two terms of (3.121) and (3.122) become zero, and we obtain the following steady-state voltage equations

$$\tilde{V}_{as} = r_s\tilde{I}_{as} + j\omega_e(L_{ls} + L_{ms})\tilde{I}_{as} + j\omega_e L_{ms}\tilde{I}'_{ar} \tag{3.123}$$

$$\tilde{V}'_{ar} = r'_r\tilde{I}'_{ar} + j(\omega_e - \omega_r)(L'_{lr} + L_{ms})\tilde{I}'_{ar} + j(\omega_e - \omega_r)L_{ms}\tilde{I}_{as} \tag{3.124}$$

The so-called "slip" is

$$s = \frac{\omega_e - \omega_r}{\omega_e} \tag{3.125}$$

If we divide (3.124) by the slip, it becomes

$$\frac{\tilde{V}'_{ar}}{s} = \frac{r'_r}{s}\tilde{I}'_{ar} + j\omega_e(L'_{lr} + L_{ms})\tilde{I}'_{ar} + j\omega_e L_{ms}\tilde{I}_{as} \tag{3.126}$$

Equations (3.123) and (3.126) suggest the single-phase equivalent T circuit of a two-phase symmetrical machine during steady-state balanced operation shown in Figure 3.19. Please note that the inductive reactances are calculated as $X = \omega_e L$. One tends to want to calculate the inductive reactances of the rotor circuit as $X = (\omega_e - \omega_r)L$ when using (3.126). It is clear that (3.124) is of the form we would expect; however, we have divided (3.124) by $(\omega_r - \omega_e)$ and multiplied by ω_e (the slip) to arrive at (3.126).

We understand that current is not induced in the rotor windings when $\omega_r = \omega_e$. Since the rotor windings are generally short circuited (v'_{ar} and v'_{br} = zero) and from (3.125) the slip is zero and $\frac{r'_r}{s}$ is infinite; hence, the rotor circuit appears to be open circuited thus correctly portraying synchronous speed "operation." We will find that, with a slight modification (X_{ms} becomes $\frac{3}{2}X_{ms}$), this circuit may also be used to calculate the per-phase steady-state performance of a three-phase symmetrical machine.

An expression for the steady-state electromagnetic torque may be obtained by first writing (3.118) in terms of I^e_{qs}, I^e_{ds}, I'^e_{qr}, and I'^e_{dr}, and then express $\tilde{I}_{as}$ and $\tilde{I}'_{ar}$ using

$$\sqrt{2}\tilde{F}_{as} = F^e_{qs} - jF^e_{ds} \tag{3.127}$$

$$\sqrt{2}\tilde{F}_{ar} = F^e_{qr} - jF^e_{dr} \tag{3.128}$$

The expression may be reduced to

$$T_e = 2\left(\frac{P}{2}\right)L_{ms}\mathrm{Re}\left[j\tilde{I}^*_{as}\tilde{I}'_{ar}\right] \tag{3.129}$$

where $\tilde{I}^*_{as}$ is the conjugate of $\tilde{I}_{as}$. The phasor currents may be calculated from the equivalent circuit given in Figure 3.19. For a three-phase machine L_{ms} must be multiplied by $\frac{3}{2}$.

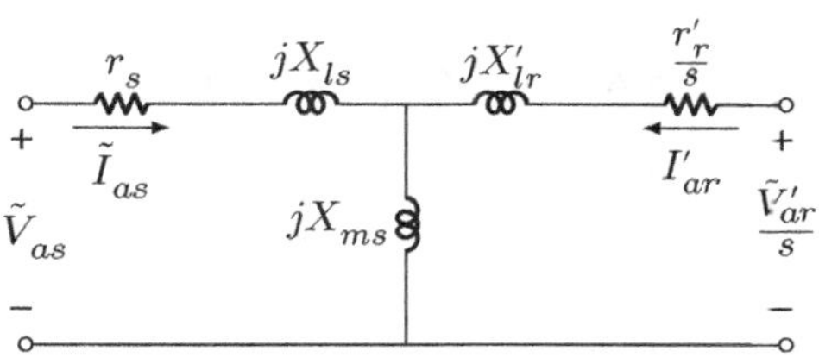

Figure 3.19 Equivalent single-phase circuit for a two-phase symmetrical induction machine for balanced steady-state operation.

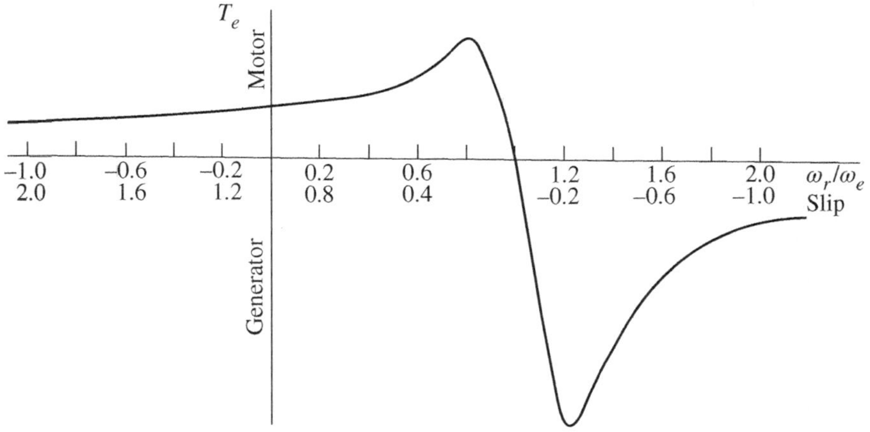

Figure 3.20 Steady-state torque-speed characteristics of a symmetrical induction machine.

Figure 3.20 shows the steady-state torque speed plot of a typical industrial-type induction motor. Stable operation occurs on the negative slope part of this plot.

Example 3F ***Phasor diagram for steady-state operation***

Consider a single-fed two-pole two-phase 5-hp 110-V (root-mean-square (rms)) 60-Hz induction machine with the following parameters: $r_s = 0.295\ \Omega$, $L_{ls} = 0.944$ mH, $L_{ms} = 35.15$ mH, $r_r' = 0.201\ \Omega$, and $L_{lr}' = 0.944$ mH. Calculate $\tilde{I}_{as}$, $\tilde{I}_{ar}'$, T_e, draw the phasor diagram and show the rotor and stator poles for (a) $s = 0.05$ and (b) $s = -0.05$. The equivalent circuit for (a) and (b) is shown in Figure 3F.1. The inductive reactances are calculated as ω_e times the inductance in henrys.

$$r_s = 0.295\ \Omega \quad X_{ms} = 13.252\ \Omega \quad r_r' = 0.201\ \Omega$$
$$X_{ls} = 0.356\ \Omega \qquad\qquad X_{lr}' = 0.356\ \Omega$$

For part (a):

$$\frac{r_r'}{s} = \frac{0.201}{0.05} = 4.02\ \Omega \tag{3F.1}$$

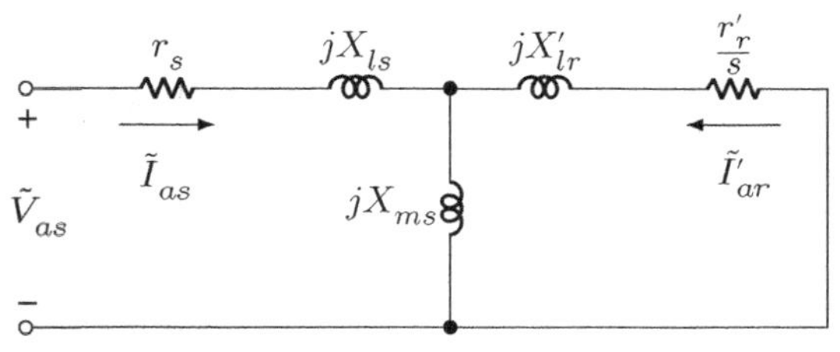

Figure 3F.1 Equivalent circuit for steady-state operation of a single-fed induction machine.

$$Z_{\text{in}} = r_s + jX_{ls} + \frac{\left(\frac{r_r'}{s} + jX_{lr}'\right) jX_{ms}}{\frac{r_r'}{s} + j(X_{lr}' + X_{ms})}$$
$$= 0.295 + j0.356 + \frac{(4.02 + j0.356)j13.252}{4.02 + j(0.356 + 13.252)}$$
$$= 4.175\underline{/24.64^\circ}\ \Omega \tag{3F.2}$$

$$\tilde{I}_{as} = \frac{\tilde{V}_{as}}{Z_{\text{in}}}$$
$$= \frac{110\underline{/0^\circ}}{4.175\underline{/24.64^\circ}} = 26.35\underline{/-24.64^\circ}\ \text{A} \tag{3F.3}$$

With $\tilde{V}_{ar} = 0$, $\tilde{I}_{ar}$ may be obtained from $\tilde{I}_{as}$ by current division. That is,

$$\tilde{I}_{ar}' = -\frac{jX_{ms}}{\frac{r_r'}{s} + jX_{rr}'}\tilde{I}_{as}$$
$$= -\frac{j13.252}{4.02 + j(0.356 + 13.252)} 26.35\underline{/-24.64^\circ}$$
$$= -24.61\underline{/-8.18^\circ} = 24.61\underline{/171.8^\circ}\ \text{A} \tag{3F.4}$$

From (3.129)

$$T_e = 2\left(\frac{P}{2}\right) L_{ms}\text{Re}\left[j\tilde{I}_{as}^*\tilde{I}_{ar}'\right]$$
$$= 2\left(\frac{2}{2}\right)(35.15 \times 10^{-3})\text{Re}[26.35\underline{/24.64^\circ + 90^\circ}\ 24.61\underline{/171.8^\circ}]$$
$$= 12.9\ \text{N m} \tag{3F.5}$$

The torque is positive in the direction of rotation; motor action. The phasor diagram for part (a) is shown in Figure 3F.2. The stator poles are "pushing" the rotor poles counterclockwise; motor action.

For part (b):

$$\frac{r_r'}{-0.05} = -4.02\ \Omega \tag{3F.6}$$

In the calculation, we need to replace $\frac{r_r'}{s} = 0.402\ \Omega$ in part (a) with $\frac{r_r'}{s} = -0.402\ \Omega$. Thus,

$$Z_{\text{in}} = 3.648\underline{/151.5^\circ}\ \Omega \tag{3F.7}$$

$$\tilde{I}_{as} = 30.154\underline{/-151.5^\circ}\ \text{A} \tag{3F.8}$$

$$\tilde{I}_{ar}' = 28.2\underline{/12^\circ}\ \text{A} \tag{3F.9}$$

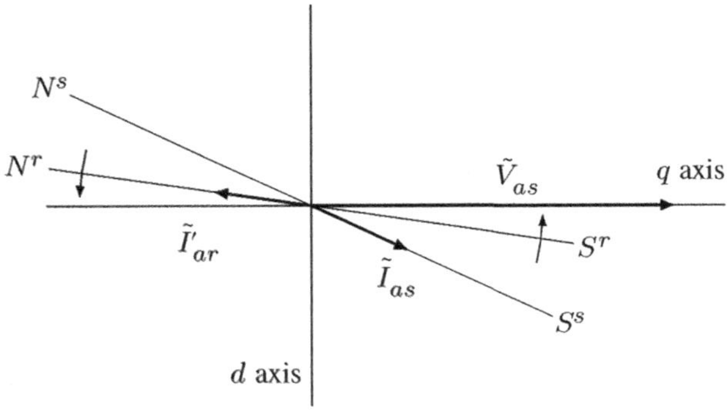

Figure 3F.2 Phasor diagram; motor action.

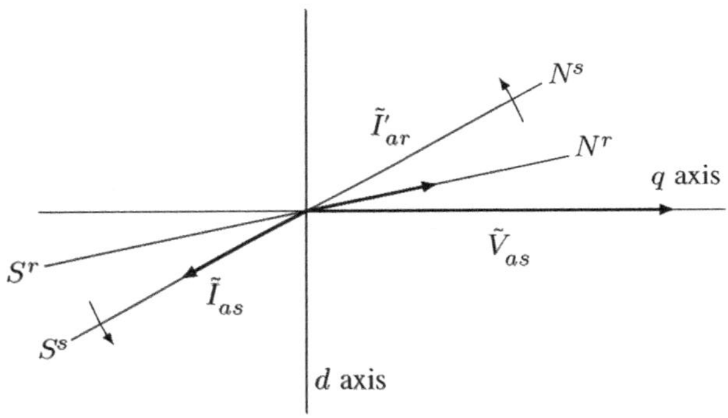

Figure 3F.3 Phasor diagram; generator action.

From (3.129)

$$T_e = 2\left(\frac{2}{2}\right)(35.15\times 10^{-3})\text{Re}[j30.15\underline{/151.5^\circ}\ 28.2\underline{/12^\circ}]$$
$$= -17.0\,\text{N m} \tag{3F.10}$$

The torque is negative in the direction of rotation; generator action. The phasor diagram is shown in Figure 3F.3. Therein, the rotor poles are "pushing" the stator poles counterclockwise; generator action.

SP3.4.1 Assume sinusoidally distributed windings on the stator and rotor of the machine shown in Figure 3.11. Express (a) mmf_{as} in terms of θ_r and ϕ_r and (b) mmf_{ar} in terms of θ_r and ϕ_s. [(a) $\text{mmf}_{as} = (N_s/2)i_{as}\cos(\phi_r + \theta_r)$; (b) $\text{mmf}_{ar} = (N_r/2)i_{ar}\cos(\phi_s - \theta_r)$]

SP3.4.2 The frequency of the balanced stator currents of an induction machine is 60 Hz, mmf_s rotates counterclockwise. The device is operating as a motor, and the rotor of the two-pole machine is rotating counterclockwise at $0.9\omega_e$. (a) Determine the frequency of the balanced rotor currents. Determine the angular velocity of mmf_s and mmf_r relative to an observer sitting (b) on the rotor and (c) on the stator. [(a) 6 Hz; (b) 37.7 rad/s, ccw; (c) 377 rad/s, ccw]

SP3.4.3 Assume that θ_r is positive in the clockwise direction in Figure 3.11 rather than in the counterclockwise direction. Express all inductances. [$L_{asbr} = L_{sr}\sin\theta_r$; $L_{bsar} = -L_{sr}\sin\theta_r$; all others unchanged]

SP3.4.4 The *as* and *bs* windings in Figure 3.11 are rotated $\frac{1}{4}\pi$ clockwise from the position shown. Express L_{asar} and L_{asbr}. $\left[L_{asar} = L_{sr}\cos\left(\theta_r + \frac{1}{4}\pi\right);\ L_{asbr} = -L_{sr}\sin\left(\theta_r + \frac{1}{4}\pi\right)\right]$

SP3.4.5 Consider the device shown in Figure 3.11. All windings are open-circuited except the *as* winding. $I_{as} = \sin t$, $L_{ms} = 0.1$ H, $L_{rr} = \frac{1}{4}L'_{rr}$, $\omega_r = 0$, and $\theta_r = \frac{1}{3}\pi$. Determine V_{ar}. [$V_{ar} = 0.025\cos t$]

SP3.4.6 In the equivalent circuits given in Figure 3.14, there are speed voltages only in the *qr* and *dr* part of the *T* equivalent circuits. What reference frame is being used? [stationary]

SP3.4.7 Repeat SP3.4.6 if speed voltages are only in the *qs* and *ds* part. [rotor]

SP3.4.8 Determine the steady-state frequency of (a) i'_{ar} and i'_{br} in Figure 3.15, (b) i'^s_{qr} and i'^s_{dr} in Figure 3.17, and (c) i^e_{qr} and i^e_{dr} in Figure 3.18. [(a) $\omega_e - \omega_r$, (b) ω_e and zero at $\omega_r = \omega_e$, (c) dc]

SP3.4.9 Repeat SP3.4.8 for the stator related variables. [(a) ω_e, (b) ω_e, (c) dc]

3.5 Problems

3.1 A permanent-magnet dc motor has the following parameters: $r_a = 8\ \Omega$ and $k_v = 0.01$ V s/rad. The shaft load torque is approximated as $T_L = K\omega_r$, where $K = 5 \times 10^{-6}$ N m s. The applied voltage is 6 V and $B_m = 0$. Calculate the steady-state rotor speed ω_r in rad/s.

3.2 A permanent-magnet dc motor is driven by a mechanical source at 3820 r/min. The measured open-circuit armature voltage is 7 V. The mechanical source is disconnected, and a 12-V electric source is connected to the armature. With zero-load torque, $I_a = 0.1$ A and $\omega_r = 650$ rad/s. Calculate k_v, B_m, and r_a.

3.3 It is found that $\lambda'_m = 0.1$ V s for a permanent-magnet six-pole two-phase ac machine. Calculate the amplitude (peak value) of the open circuit phase voltage measured when the rotor is turned at 60 revolutions per second (r/s).

3.4 A four-pole two-phase permanent-magnet ac machine is driven by a mechanical source at $\omega_r = 3600$ r/min. The open-circuit voltage across one of the phases is 50 V (rms). (a) Calculate λ'_m. The mechanical source is removed and the following voltages are applied: $V_{as} = \sqrt{2}\,25\cos\theta_r$, $V_{bs} = \sqrt{2}\,25\sin\theta_r$, where $\theta_r = \omega_r t$. (b) Neglect friction ($B_m = 0$) and calculate the no-load rotor speed ω_r in rad/s.

3.5 The parameters of a two-pole two-phase permanent magnet ac machine are $r_s = 4\ \Omega$, $\lambda'^r_m = 0.07$ V s/rad, $L_{ls} = 1$ mH, $L_{ms} = 9$ mH. The applied voltages are $V_{as}\sqrt{2}\,14\cos\theta_r$ and $V_{bs} = \sqrt{2}\,14\sin\theta_r$ where $\theta_r = 200t$. Determine (a) $\tilde{I}_{as}$, (b) T_e, and (c) show the phasor diagram and locate the poles.

3.6 A four-pole two-phase induction machine has the following parameters: $r_s = 0.3\ \Omega$, $L_{ls} = 1$ mH, $L_{ms} = 20$ mH, $r'_r = 0.2\ \Omega$, and $L'_{lr} = 1$ mH. The device is supplied from a 60-Hz source; the rotor speed is $\omega_r = 360$ rad/s. In this mode of operation $\tilde{I}_{as} = 28.8\underline{/-36.1^\circ}$ and $\tilde{I}_{ar} = 23.9\underline{/173.2^\circ}$. Calculate (a) T_e, (b) the total ohmic loss in the rotor windings, (c) the mechanical power delivered to the load, and (d) express I_{as}, I'_{ar}, I^S_{qs}, and I'^S_{qr}.

3.7 Consider Problem 3.6. Calculate V_{as}. Draw the phasor diagram and locate the poles. Show all voltage drops associated with the equivalent circuit shown in Figure 3F.1.

3.8 Select two identical capacitors so that when they are connected in parallel with each phase (one capacitor per phase) of the induction machine described in Example 3F, the no-load capacitor–induction machine combination operates at unity power factor. Assume that capacitors are ideal (zero resistance).

3.9 Obtain (3.126).

References

1 Krause, P.C. (1986). *Analysis of Electric Machinery*. McGraw-Hill.
2 Krause, P.C., Wasynczuk, O., Sudhoff, S.D., and Pekarek, S.D. (2013). *Analysis of Electric Machinery and Drive Systems*, 3e. IEEE Press/Wiley.
3 Krause, P.C. and Wasynczuk, O. (1989). *Electromechanical Motion Devices*. McGraw-Hill.
4 Krause, P.C., Wasynczuk, O., Pekarek, S.D., and O'Connell, T.C. (2019). *Electromechanical Motion Devices, "Rotating Magnetic Field Based Analysis with Online Animations"*, 3e. IEEE Press/Wiley.
5 Krause, P.C., Wasynczuk, O., O'Connell, T.C., and Hasan, M. (2017). *Introduction to Electric Power and Drive Systems*. IEEE Press/Wiley.

4

Power Electronics

4.1 Introduction

Power electronic circuits are the foundation of electric drives. Power-electronic-drive-controlled machines enable propulsion systems in air-, marine-, and ground-based vehicles, aircraft actuation and industrial systems, and many others. This chapter lays the groundwork for advanced study in these areas, first presenting the fundamental principles of switching circuits and then analyzing some of the most common power-electronic circuits, focusing on those used in electric drives.

In Section 4.2, we introduce several fundamental concepts relevant to power conversion that can be applied across a broad range of basic and advanced circuits. In Sections 4.3–4.5, these concepts are used to introduce dc–dc (direct-current) converters, ac–dc rectifiers, and dc–ac (alternating-current) inverters, respectively. This introduction includes the common buck and boost dc–dc converters, half- and full-wave rectifiers, and the single-phase inverter. The three-phase inverter (also called the "six-step" inverter) is covered in Chapter 5 as part of a brushless dc drive.

ac–ac conversion, the logical successor to these topics, is conspicuously absent. Why? As shown in Chapter 1, ac transformers are ideal for ac–ac conversion *if* the transformation is at a fixed frequency. Since this is usually the case, power electronic converters are not as prevalent for this application as they are for the others covered here. But when the frequency must be changed, a transformer is insufficient and an ac–ac power-electronic circuit – the cycloconverter is one example – is required. Cycloconverters and other ac–ac conversion methods are left as topics for advanced study.

Introduction to the Analysis of Electromechanical Systems, First Edition.
Paul C. Krause, Oleg Wasynczuk, and Timothy O'Connell.

4.2 Switching-circuit Fundamentals

The term "power electronics" encompasses electronic circuits and devices that control electric energy flow. Power electronic devices are ubiquitous, operating diligently behind the scenes to power an ever-expanding array of electronic devices. Although most power converters are designed to be inconspicuous, their essential role in today's world cannot be overstated. Because electrical energy is produced and consumed over a wide range of voltages and frequencies, the applications for power electronic devices are truly limitless.

4.2.1 Power Conversion Principles

The notional power electronic circuit shown in Figure 4.1 provides an interface between two circuits, facilitating the flow of power between them when a direct connection is inefficient or impossible. For example, a dc–ac inverter can act as an interface between a dc battery and an ac electric motor.

The connected circuits denoted "Circuit 1" and "Circuit 2" in Figure 4.1 traditionally have been characterized as *source* and *load*, respectively, where a power-producing circuit is a source and a power-consuming circuit a load. Modern trends in the power field have blurred the line between sources and loads, as many devices are both producers and consumers of power. For example, the electric machine in a hybrid electric vehicle can act as both a load (motor mode) and source (generator mode) depending on the state of operation of the vehicle. Power electronic circuits can, and often do, facilitate bidirectional power flow between circuits, as indicated by the double arrows in Figure 4.1.

There are two key principles of power conversion [1]. First, power conversion circuits should consume as little power as possible. An ideal power device is one that is 100% efficient and consumes none of the power flowing through it. Second, power conversion circuits should transfer/process much more power than they consume. This allows a relatively small power device to interface between two devices that process large amounts of power. For example, a reasonably efficient 10 kW inverter might be 97% efficient. Thus, while the inverter can transfer 10 kW from dc source to ac load, it only consumes 300 W in the process.

Power conversion circuits often interface between dc systems operating at different voltages. For this application, a naïve circuit choice is the resistive voltage divider. Although this circuit is both simple and a great choice for logic-level

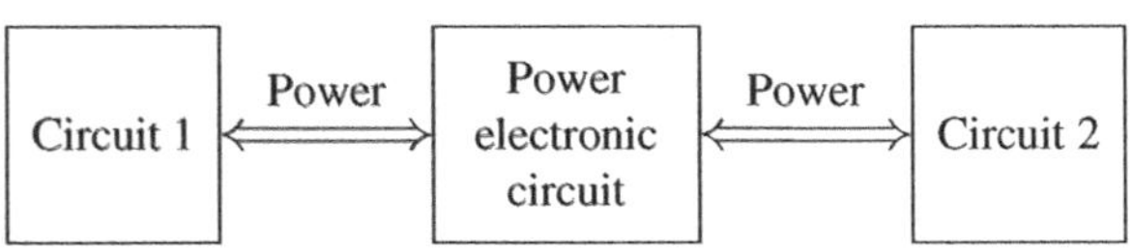

Figure 4.1 A generic power-electronic circuit.

signals, it violates both power conversion principles because it is so inefficient. Thus, a voltage divider is not a power-electronic circuit. A better choice is the dc–dc converter, which is detailed in Section 4.3. While it would seem that this application is also well suited to the function provided by a transformer, unfortunately, a transformer only operates with ac voltages. DC voltage leads to saturation of the transformer's magnetic core, rendering it useless for power transfer. Because of this, achieving a so-called "dc transformer action" is one of the core applications of power electronics.

4.2.2 Switches and Switching Functions

In power electronics, a *switch* is a device that regulates electric power flow. Every power-electronic circuit has at least one switch, and it is the operation and location of its switches that determine the circuit's behavior. A switch can be controlled (e.g. a transistor) or uncontrolled (e.g. a diode). A transistor found in power circuits is typically a bipolar junction transistor (BJT), metal-oxide-semiconductor field-effect transistor (MOSFET) or insulated-gate bipolar transistor (IGBT).

The IGBT has become popular for electric drives because of its fast switching speeds, simple voltage-controlled switching and high power, and current capabilities. It is the industrial workhorse device. However, new wide bandgap silicon-carbide (SiC) and gallium-nitride (GaN) MOSFETs, which can switch much faster than traditional silicon-based devices, are becoming more competitive with IGBTs as their voltage and power ratings and reliability increase. Diodes can be p–n junction or metal-oxide-semiconductor (MOS) devices. Although they are physically less complex than transistors, diodes vary widely in their current-carrying and voltage-blocking capabilities depending on their construction.

A real switch can be approximated as an *ideal switch*, a theoretical concept whose circuit representation is shown in Figure 4.2. An ideal switch is lossless: it carries no current and maintains an open circuit between its terminals when it is open/off and has no voltage drop and maintains a short circuit between its terminals when it is closed/on. Real switching devices can be approximated as ideal to the first order, and much meaningful analysis comes from this. Thus, to avoid unnecessary complication, we limit our discussion and subsequent analyses to ideal switches.

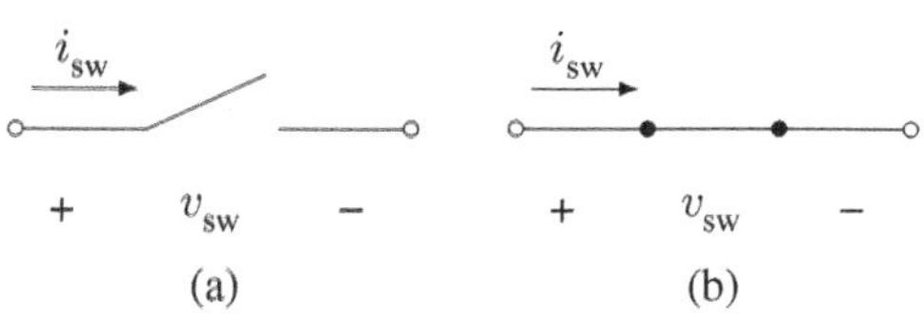

Figure 4.2 Ideal switch in (a) open/off and (b) closed/on configurations.

The ideal switch voltage, v_{sw}, and current, i_{sw}, behaviors are summarized as follows:

$$\begin{aligned} &\text{Switch open} \Rightarrow i_{sw} = 0, \\ &\text{Switch closed} \Rightarrow v_{sw} = 0 \end{aligned} \tag{4.1}$$

This behavior is described mathematically using a binary switch variable, q, as

$$q = \begin{cases} 0, & \text{switch open} \\ 1, & \text{switch closed} \end{cases} \tag{4.2}$$

Note that q here is a variable and is different from the label of the q axis introduced in Chapter 2. To avoid confusion, we could use a different variable name here; however, in an attempt at consistency with traditional power-electronic circuit analysis notation, we will not. Through the remainder of the text, it should be clear from context which q is being used.

A power-electronic switch does not remain in one state for long; rather, it is opened and closed rapidly – at kHz to MHz frequencies – to transfer power effectively. The switch variable q can thus be described with a transient *switching function*, $q(t)$, which opens and closes the switch at a fixed rate known as the *switching frequency*, f_{sw}. The inverse of the switching frequency is the *switching period*, $T_{sw} = 1/f_{sw}$. The fraction of the switching period during which the switch is on is the *duty ratio*, k. The switching frequency is chosen based on many factors, among these are efficiency, circuit ripple and/or distortion requirements, acoustic noise, and device ratings.

A common switching function encountered in the design of power-electronic converters is shown in Figure 4.3. The *Fourier series* representation of this generalized switching function is

$$q(t) = k + \frac{2}{\pi}\sum_{n=1}^{\infty}\frac{\sin(n\pi k)}{n}\cos(n\omega_{sw}t - n\theta(0)) \tag{4.3}$$

where $\omega_{sw} = 2\pi f_{sw} = 2\pi/T_{sw}$ is the frequency of the fundamental ($n = 1$) component and $\theta(0)$ is its phase shift angle at time zero. The switching function's average value is its duty ratio, k. Note that the square wave in Figure 4.3 is an

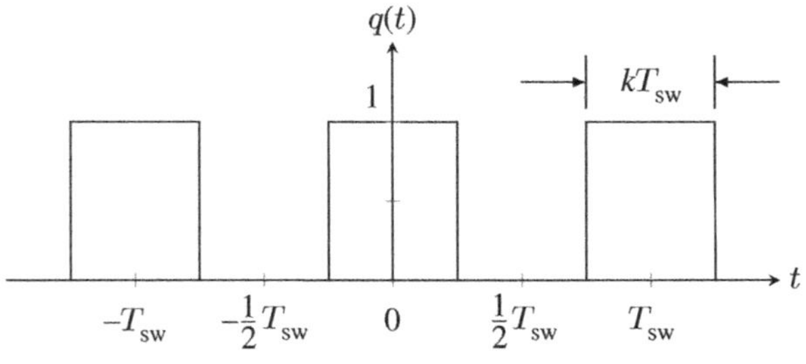

Figure 4.3 General periodic switching function $q(t)$.

even function centered at $t = 0$, which means that its fundamental has $\theta(0) = 0$. However, the Fourier series in (4.3) is general and can easily be adjusted for shifted square waves by setting $\theta(0)$ appropriately. Importantly, (4.3) reveals that a periodic signal with a dc component can have its harmonics (i.e. its sinusoidal series terms) filtered out to give a constant value dc signal. In other words, the general periodic switching function $q(t)$ in Figure 4.3 can be filtered to give a dc signal with constant value k. This is a key concept that is utilized extensively in power conversion.

While switching functions are used to control switches, uncontrolled devices, such as diodes, are not actively controlled. Rather, ideal diodes obey the following:

$$\begin{aligned} &\text{Diode off}: \quad i_D = 0 \Rightarrow v_D < 0 \\ &\text{Diode on}: \quad v_D = 0 \Rightarrow i_D > 0 \end{aligned} \tag{4.4}$$

where v_D and i_D are the diode voltage and current, respectively, as shown in Figure 4.4. In power-electronic circuits, controlled and uncontrolled switches are often used in tandem, so that if the controlled switch is opened and closed according to $q(t)$, the uncontrolled switch will obey $1 - q(t)$. When one switch is open, the other is closed, and vice versa. An uncontrolled switch is simpler to use and often cheaper than a similarly rated controlled device; however, a controlled device that follows a $1 - q(t)$ switching function can be used to perform the same function. This is often done in high-performance circuits, where precise control of all switches is important.

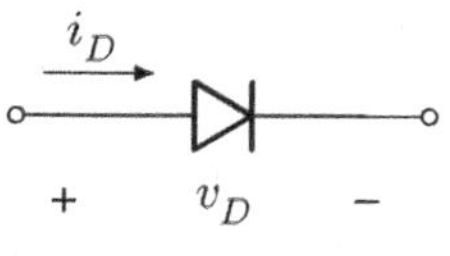

Figure 4.4 Ideal diode circuit representation.

The ideal switch concept approximates reality. Although sufficient for our introductory treatment, the switching rules in (4.1) are not strictly true for real devices. In fact, IGBTs, MOSFETs, diodes, and other real switches have finite turn-on and turn-off transition times. During these transitions, which are typically less than 1 μs, both v_{sw} and i_{sw} are non-zero concurrently as they transition to their final state. Because $v_{sw} \times i_{sw} \neq 0$, power is lost in the switch. This *switching loss* is proportional to f_{sw} and negatively impacts a power converter's efficiency. In addition, due to both solid-state physical effects and ohmic losses, the voltage across a real switch is non-zero when the switch is fully on and conducting current. This *conduction loss*, which also negatively impacts a converter's efficiency, is proportional to duty cycle – the longer a switch is on, the higher its conduction loss. Because they affect efficiency and generate heat, switching and conduction losses are important considerations in the design of high-performance power converters. For a thorough treatment of this topic, interested students should consult a dedicated, comprehensive power electronics text, such as [1–3].

Example 4A ***Fourier series of a switching function***

Let us now calculate the Fourier series of the switching function given in Figure 4A.1. Note that the frequency of the function in this illustrative example is much smaller than would be found in a typical power converter.

We must compute the Fourier series parameters in (4.3) by comparing Figure 4A.1 to Figure 4.3. By inspection, we see that

$$T_{sw} = 2\pi \tag{4A.1}$$

$$k = \frac{\pi/2}{T_{sw}} = \frac{\pi/2}{2\pi} = 1/4 \tag{4A.2}$$

and

$$\omega_{sw} = 2\pi/T_{sw} = 1 \text{ rad/s} \tag{4A.3}$$

The second pulse in the example is centered at $t = \pi/4$, as indicated by the dotted vertical line. This is 1/8 of the switching period. Thus,

$$\begin{aligned} \theta(0) &= \omega_{sw} \times \frac{T_{sw}}{8} \\ &= (1 \text{ rad/s}) \times \left(\frac{2\pi}{8}\text{s}\right) = \pi/4 \text{ rad} \end{aligned} \tag{4A.4}$$

We can now write the Fourier series using (4A.1)–(4A.4) and (4.3):

$$q(t) = \frac{1}{4} + \frac{2}{\pi}\sum_{n=1}^{\infty} \frac{\sin(n\pi/4)}{n} \cos(nt - n\pi/4) \tag{4A.5}$$

We note that the Fourier series terms (i.e. the *harmonics*) that are multiples of $1/k = 4$ are zero because of the sine coefficient term. This relationship – harmonics that are multiples of the reciprocal of the duty ratio are zero – is a general one, and it is important in advanced power-electronic circuit analysis and design.

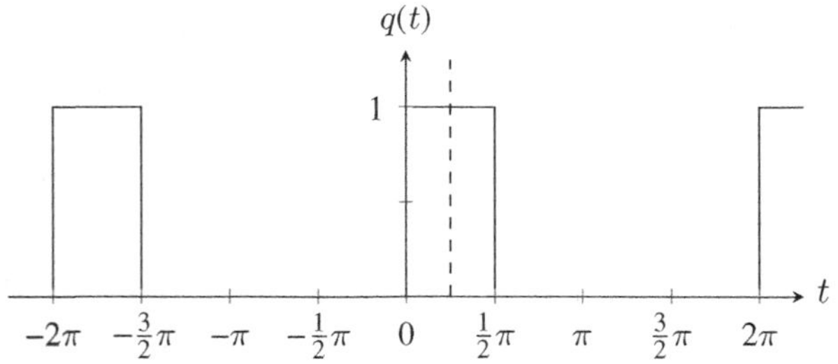

Figure 4A.1 Switching function.

4.2.3 Energy Storage Elements

Power circuits require inductors and/or capacitors, which act as *energy storage elements*. Most power-electronic circuits are constructed with components from just three categories: switches, inductors (possibly coupled), and capacitors. Two important design concepts, based on the principle of conservation of energy, govern the use of energy storage devices in power circuits.

It is instructive to consider the switched inductor circuit shown in Figure 4.5a. When the switch is closed, the inductor's current will ramp up linearly according to $di/dt = V_{\text{dc}}/L$ and the energy stored in its magnetic field, $1/2Li^2$, will increase quadratically. Theoretically, if the switch remains closed, the current will ramp up indefinitely and the inductor will continue to store energy at an ever-increasing rate. Of course, in a real inductor its conductors will overheat and melt before the current gets too high, but the takeaway is the same: it is a bad idea to close a dc source into an inductor indefinitely. It is better to maintain a constant average current and transfer energy into and out of the inductor cyclically. In the circuit shown, this would require additional circuitry, but we'll leave that detail out for now. If the energy stored during charging equals the energy extracted during discharging, the inductor will not overheat. Simple analysis (see Example 4B) shows that this steady-state operation is achieved if and only if the average inductor voltage, $\hat{v}_L$, is zero. Therefore, the first key design principle is

$$\hat{v}_L = 0 \tag{4.5}$$

Analogous arguments can be applied to the switched capacitor circuit shown in Figure 4.5b. When the switch is opened, the capacitor's voltage will increase linearly according to $dv/dt = I_{\text{dc}}/C$ and the energy stored in its electric field, $1/2Cv^2$, will increase quadratically. Theoretically, the voltage will ramp up indefinitely and the capacitor will continue to store energy at an ever-increasing rate. Of course, as with a real inductor, in a real capacitor physical reality will kick in; the voltage will eventually reach the breakdown voltage of the capacitor separator and arc across the capacitor terminals, creating a short circuit. Thus, it is a bad idea to close a

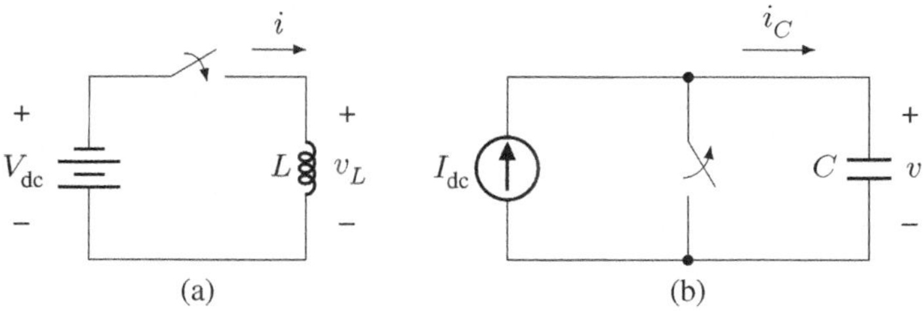

Figure 4.5 Switched (a) inductor and (b) capacitor circuits.

dc current source into a capacitor indefinitely. Instead, energy should be transferred into and out of the capacitor periodically, so that the average voltage remains constant. If the energy stored during charging equals the energy extracted during discharge, the capacitor will not overcharge and short circuit. One can show that this steady-state operation can be achieved if and only if the average capacitor current, $\hat{i}_C$, is zero. Thus, the second key design principle is

$$\hat{i}_C = 0 \tag{4.6}$$

We now have the necessary tools to begin an introductory analysis of several power electronic switching circuits. Together with basic circuit theory, these enable us to analyze dc–dc, ac–dc, and dc–ac circuits in the following sections.

Example 4B ***Inductor average voltage for energy balance***
This example demonstrates that the average inductor voltage must be zero for energy balance in the switched inductor circuit in Figure 4.5a.

The inductor voltage, v_L, is expressed

$$v_L = L\frac{di}{dt} \tag{4B.1}$$

With constant inductance, the average inductor voltage is

$$\hat{v}_L = L\widehat{\frac{di}{dt}} \tag{4B.2}$$

If $\widehat{\frac{di}{dt}} \neq 0$, then the inductor current magnitude will continually increase. Since the stored inductor energy is $1/2Li^2$, this represents an unbalanced energy condition. Therefore, for energy balance

$$\widehat{\frac{di}{dt}} = 0 \tag{4B.3}$$

Combining (4B.2) and (4B.3) for energy balance reveals that

$$\hat{v}_L = 0 \tag{4B.4}$$

SP4.2.1 (a) Find the voltage conversion efficiency $\eta = P_{out}/P_{in}$ for a voltage divider circuit with $r_1 = 5\,\Omega$, $r_2 = 5\,\Omega$, and $r_{out} = 100\,\Omega$ (r_1 is in series with the parallel combination of r_2 and r_{out}. You may refer to Figure 4.39 in the chapter problems). (b) Is this a good power conversion circuit? [(a) $\eta = 0.023 = 2.3\%$; (b) no, approximately 98% of the input power is consumed in the circuit!]

SP4.2.2 Show that $\hat{i}_C = 0$ is necessary for energy balance in a capacitor. [See Example 4B and use analogous arguments]

4.3 dc–dc Conversion

Let us consider the ideal dc–dc conversion circuit conceptualized in Figure 4.6. Power is to be transferred from the dc source, V_{hi}, to the lower voltage dc load, V_{lo}. Although this motivating example transfers power from a higher voltage source to a lower voltage load, the analysis concepts that follow are equally applicable for low-to-high voltage conversion circuits. From examining the Fourier series in (4.3), we have seen that a periodic signal with a dc component can be filtered to give a constant value dc signal. Thus, if we can create a periodic signal from V_{hi} that has an average value of V_{lo}, we can filter out the harmonic content and be left with the desired output. A simple periodic signal that meets the requirements is a square wave with magnitude V_{hi} and duty ratio $k = V_{lo}/V_{hi}$. k has a value between 0 and 1, and the length of time, per period, that the square wave is high is kT_{sw}, as is indicated in Figure 4.7.

This suggests a dc–dc conversion strategy: transfer power from V_{hi} in square pulses with $k = V_{lo}/V_{hi}$ and filter the result to leave only the dc component, V_{lo}. Because a low-pass filter is required to eliminate the ac content, the higher the frequency of the pulses, the simpler and more effective this filter can be.

4.3.1 Buck Converter

A circuit that accomplishes the proposed dc–dc conversion strategy is the *buck converter*, shown in Figure 4.8. This circuit has one controlled switch – a transistor – and one uncontrolled switch – a diode. Assuming L and C are "large" (we will see what this means a little later) and that the circuit is in steady state, we

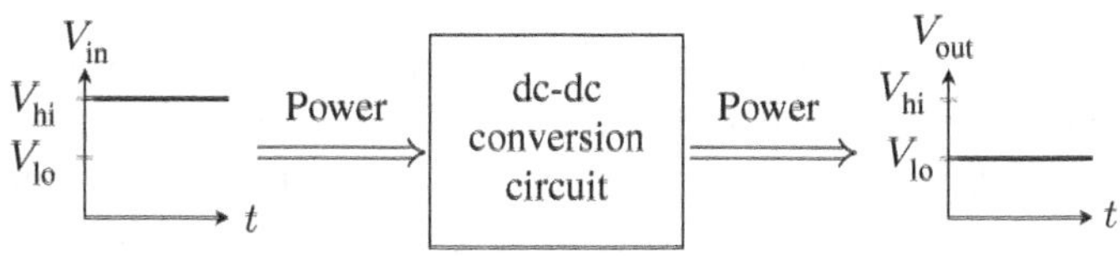

Figure 4.6 Ideal dc–dc conversion circuit.

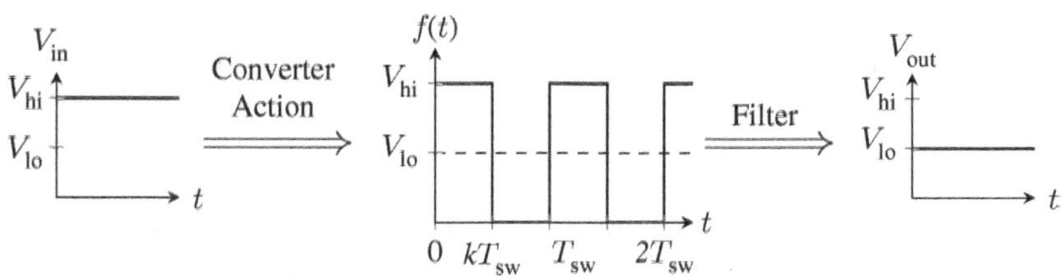

Figure 4.7 Filtered square wave voltage dc–dc voltage conversion strategy.

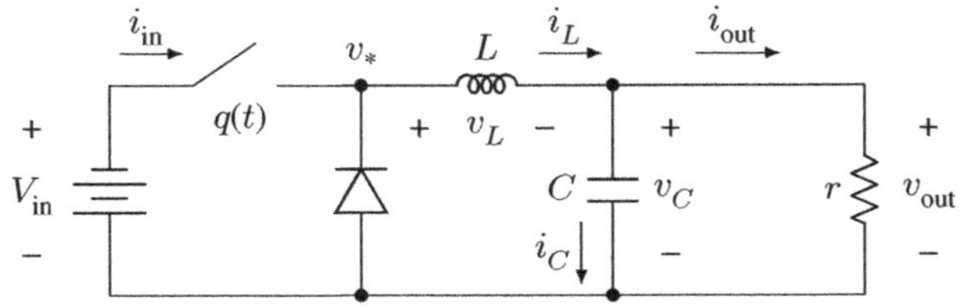

Figure 4.8 Buck converter circuit.

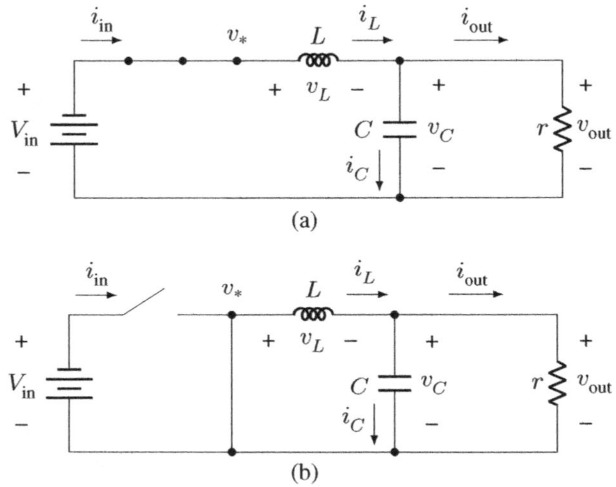

Figure 4.9 Buck converter in (a) Configuration A, with $q(t) = 1$, and (b) Configuration B, with $q(t) = 0$.

can approximate as constants the output voltage, v_{out}, and the inductor current, i_L. When the transistor is closed, the diode is reverse-biased and does not conduct current; the circuit is in Configuration A, as shown in Figure 4.9, and current flows through the inductor from the input to the output. When the transistor opens, the inductor current cannot change instantaneously and Kirchhoff's current law (KCL) requires that this current have a closed path in which to flow. As the inductor's stored energy is dissipated in r, its current decreases, inducing a negative inductor voltage, v_L. This forward-biases the diode, turning it on and providing a path for i_L; the circuit switches to Configuration B. An appropriate switching function controls the transistor, switching the circuit between these two configurations.

Let us now analyze Configurations A and B in more detail. In Configuration A, $i_{\text{in}} = i_L$, $v_* = V_{\text{in}}$, and $v_L = V_{\text{in}} - v_{\text{out}}$. In Configuration B, the input voltage is detached from the circuit, $v_* = 0$, $i_{in} = 0$, and $v_L = -v_{out}$. If we assume that the transistor is switched at a high rate so that the circuit rapidly oscillates between Configurations A and B, an average behavior will emerge. We will first analyze this, then add the high-frequency details. In terms of the switching function,

$$\begin{aligned} v_*(t) &= q(t)V_{in} + (1 - q(t)) \times 0 \\ &= q(t)V_{in} \end{aligned} \tag{4.7}$$

The square wave voltage, v_*, can be described using (4.3) with $\theta(0) = k\pi$. The time average of v_* over one switching period is

$$\begin{aligned} \hat{v}_* &= \frac{1}{T_{sw}} \int_0^{T_{sw}} v_* dt \\ &= \frac{1}{T_{sw}} \left(\int_0^{kT_{sw}} V_{in} dt + \int_{kT_{sw}}^{T_{sw}} 0 \times dt \right) \\ &= \frac{1}{T_{sw}} \int_0^{kT_{sw}} V_{in} dt = kV_{in} \end{aligned} \tag{4.8}$$

Now, because the average voltage across the inductor is 0 (4.5), we have found the average output voltage:

$$\hat{v}_{out} = \hat{v}_* = kV_{in} \tag{4.9}$$

The buck converter's voltage *conversion ratio* is then

$$\frac{\hat{v}_{out}}{V_{in}} = k \tag{4.10}$$

Equations (4.9) and (4.10) equivalently describe the average behavior of the buck converter, showing that the average output voltage is a fraction of the input voltage and is determined by the duty ratio k. Look again at Figure 4.7. The equation we have derived encompasses all the features of that figure. The input voltage is first "chopped" by the controlled switch into the square wave v_*. A buck converter is sometimes called a "*chopper*" because of this. With rapid enough switching and "large" L and C, v_* is filtered to leave its average value as kV_{in}. Thus, $\hat{v}_{out}/V_{in} = k$ and dc voltage transformation has been accomplished!

Let us now consider the current and power flows in this circuit. Because the average current in the output capacitor is zero (4.6),

$$\hat{i}_{out} = \hat{i}_L \tag{4.11}$$

Further, the input current can be expressed in terms of the switching function as

$$\begin{aligned} i_{in}(t) &= q(t)i_L + (1 - q(t)) \times 0 \\ &= q(t)i_L \end{aligned} \tag{4.12}$$

The time-average input current over one switching period is

$$\begin{aligned}\hat{i}_{\text{in}} &= \frac{1}{T_{\text{sw}}}\int_0^{T_{\text{sw}}} i_{\text{in}}(t)dt \\ &= \frac{1}{T_{\text{sw}}}\left(\int_0^{kT_{\text{sw}}} i_L dt + \int_{kT_{\text{sw}}}^{T_{\text{sw}}} 0 \times dt\right) \\ &= \frac{1}{T_{\text{sw}}}\int_0^{kT_{\text{sw}}} i_L dt = k\hat{i}_L \end{aligned} \tag{4.13}$$

In the last step we have used our assumption that i_L is approximately constant, and we have set $i_L = \hat{i}_L$. The average power into the converter, $\hat{P}_{\text{in}}$, is

$$\hat{P}_{\text{in}} = V_{\text{in}}\hat{i}_{\text{in}} = V_{\text{in}}k\hat{i}_L \tag{4.14}$$

and the average power out of the converter, $\hat{P}_{\text{out}}$, is approximated as

$$\hat{P}_{\text{out}} = \widehat{(v_{\text{out}}i_{\text{out}})} = \hat{v}_{\text{out}}\hat{i}_L \tag{4.15}$$

To set the average of a product equal to the product of two averages, which is in general not a valid mathematical step, we have used our assumption that v_{out} and i_L are constant and equal to $\hat{v}_{\text{out}}$ and $\hat{i}_L$, respectively. Under this assumption (4.15) is valid. Substituting the buck converter voltage relationship in (4.9) into (4.14) and comparing to (4.15) yields

$$\hat{P}_{\text{in}} = \hat{v}_{\text{out}}\hat{i}_L = \hat{P}_{\text{out}} \tag{4.16}$$

showing that the average power is conserved. The ideal buck converter transfers all the average input power to the output and consumes none, making it an exemplary power conversion device.

Thus far, we have left out a few important details. Namely: what constitutes "large" L and C, and what is "fast" switching? We will look at these details now; however, before we do it is worth reviewing the power conversion strategy embodied in Figure 4.7 and the discussion and equations leading to (4.9). Similar strategies and concepts are used over and over in power electronic converters, and a firm understanding now will go a long way toward the understanding of more advanced devices.

Let us now analyze the steady-state operation of the buck converter circuit. We will examine the inductor current and capacitor voltage more closely to gain insight into the relationship between L, C, and f_{sw}. Thus far, we have assumed constant v_{out} and i_L. To analyze the inductor current in more detail, we will continue to assume that the output capacitance is large so that the capacitor can be approximated as a dc voltage source with voltage V_{out}. Figure 4.10 shows the approximate circuit.

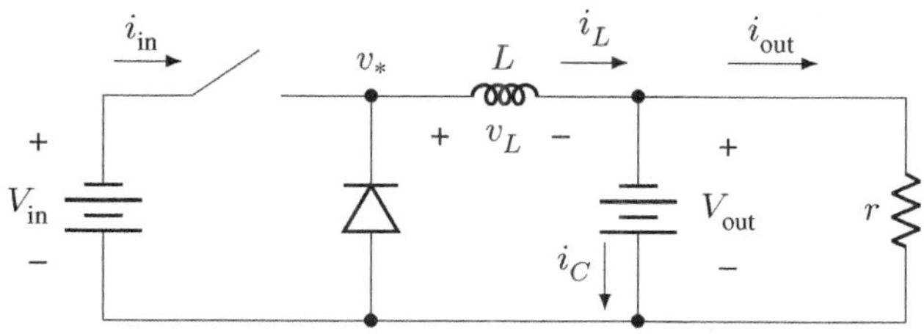

Figure 4.10 Buck converter with "large" output capacitance approximated as a voltage source.

The inductor voltage can be expressed using (4.7) as

$$\begin{aligned} v_L &= v_* - V_{\text{out}} \\ &= q(t)V_{\text{in}} - V_{\text{out}} \end{aligned} \tag{4.17}$$

Because the inductor voltage is also $L di_L/dt$, we have

$$\frac{di_L}{dt} = \frac{q(t)V_{\text{in}} - V_{\text{out}}}{L} \tag{4.18}$$

When the transistor is on and the circuit is in Configuration A, the inductor current ramps up linearly with slope $(V_{\text{in}} - V_{\text{out}})/L$; when the transistor is off and the circuit is in Configuration B, the current ramps down linearly with slope $-V_{\text{out}}/L$. The triangular-shaped periodic waveform with average $\hat{i}_L = I_L$ shown in Figure 4.11 is the result. The peak-to-peak ripple magnitude of this waveform is ΔI, and its period is T_{sw}.

We can calculate ΔI by taking the magnitude of the result of integrating (4.18) with respect to t either from 0 to kT_{sw} or from kT_{sw} to T_{sw}. In either case, the result is

$$\begin{aligned} \Delta I &= \frac{V_{\text{out}}}{L}(T_{\text{sw}} - kT_{\text{sw}}) \\ &= \frac{V_{\text{out}}}{L \times f_{\text{sw}}}(1-k) \end{aligned} \tag{4.19}$$

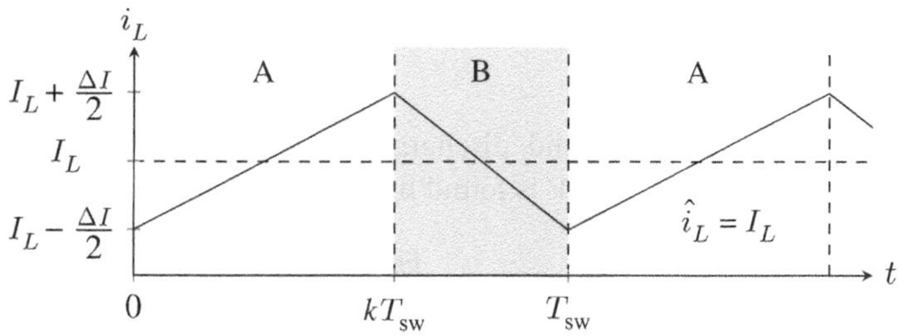

Figure 4.11 Buck converter inductor current waveform over one-and-a-half switching periods.

Notice that the ripple magnitude is a function of L and f_{sw}. For a given switching frequency, if a maximum ripple ΔI_{max} is specified, the minimum inductance necessary is

$$L_{min} = \frac{V_{out}(1-k)}{f_{sw} \times \Delta I_{max}} \tag{4.20}$$

According to (4.20), if $L \geq L_{min}$, then $\Delta I \leq \Delta I_{max}$.

We have thus derived a minimum threshold for determining a "large" L, based on the switching frequency and a specified maximum allowable ripple current. But, what about C? Thus far, C has been assumed large so that the capacitor could be considered a dc voltage source. In reality, the capacitor voltage will have an average value with some triangular ripple, similar to the inductor current waveform. To analyze the capacitor, we will use the circuit shown in Figure 4.12. The inductor current is represented as a current source with average $\hat{i}_L = I_L$ and triangular ripple of magnitude $\pm\Delta I/2$. The capacitor is in parallel with the resistive load. Because according to (4.6), the capacitor has no average current in the steady state, I_L flows through the resistor and the ripple current flows through the capacitor. In reality, a small fraction of the ripple current will also flow through the resistive load; however, if the capacitor impedance at the ripple frequency is much smaller than the load impedance, which it will be in normal circumstances, then the majority of the ripple current will flow through the capacitor and our assumption is valid. In fact, in many instances, with a reasonable amount of output capacitance, the inductance in the conductors connecting the converter to the load is large enough to validate our assumption, regardless of the load impedance.

The capacitor voltage, v_C, will rise and fall according to

$$\frac{dv_C}{dt} = \frac{i_C(t)}{C} \tag{4.21}$$

where i_C is the ripple current from the inductor. The capacitor current is shown in Figure 4.13, which is similar to Figure 4.11 except that the dc offset has been removed.

The variable δ_{mid} is introduced to mark the midway point between kT_{sw} and T_{sw}:

$$\delta_{mid} \triangleq \frac{T_{sw} - kT_{sw}}{2} \tag{4.22}$$

The capacitor charges when $i_C > 0$ and discharges when $i_C < 0$. Thus, the capacitor peak-to-peak voltage ripple ΔV is found by calculating the voltage rise

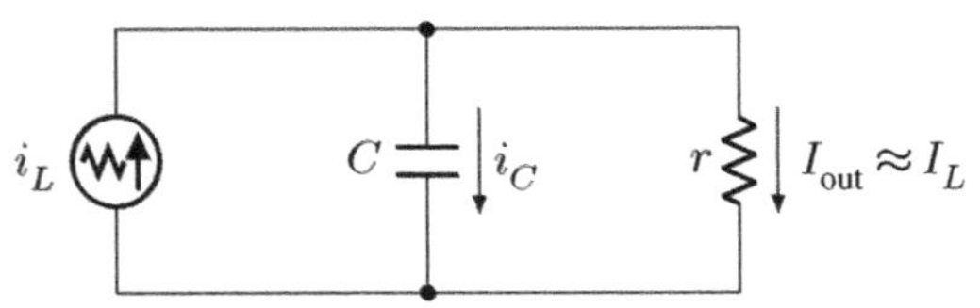

Figure 4.12 Buck converter output stage used to calculate C_{min}.

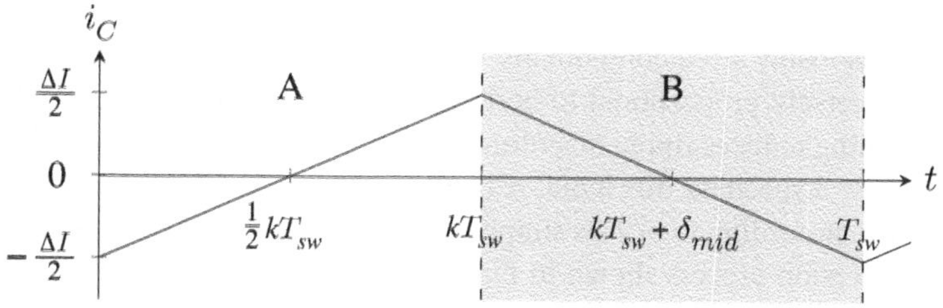

Figure 4.13 Buck converter approximate capacitor current.

that occurs between $kT_{sw}/2$ and $kT_{sw} + \delta_{mid}$. Because the ripple is symmetrical about the t-axis, its magnitude can also be found by calculating the voltage drop that occurs when i_C is negative; however, because this requires integrating over a split interval, we choose the positive region for simplicity. Specifically,

$$\begin{aligned}\Delta V &= \int_{kT_{sw}/2}^{kT_{sw}+\delta_{mid}} \frac{dv_C}{dt}dt \\ &= \frac{1}{C}\frac{\left[\left(\frac{kT_{sw}}{2}\times\frac{\Delta I}{2}\right)+\left(\delta_{mid}\times\frac{\Delta I}{2}\right)\right]}{2} \\ &= \frac{\Delta I}{8C}(kT_{sw}+2\delta_{mid})\end{aligned} \tag{4.23}$$

Using (4.22) to simplify, this becomes

$$\begin{aligned}\Delta V &= \frac{\Delta I\times T_{sw}}{8C} \\ &= \frac{\Delta I}{8C\times f_{sw}}\end{aligned} \tag{4.24}$$

The voltage ripple magnitude is a function of C, f_{sw}, and the current ripple magnitude. For a given switching frequency and current ripple, then, the minimum capacitance C_{min} necessary to keep the output voltage ripple below a desired limit ΔV_{max} is

$$C_{min} = \frac{\Delta I}{8f_{sw}\Delta V_{max}} \tag{4.25}$$

According to (4.25), if $C \geq C_{min}$, then $\Delta V \leq \Delta V_{max}$. Thus, we have derived a minimum threshold for determining a "large" C, based on the switching frequency, the allowable voltage ripple, and the inductor current ripple.

The last detail is the switching frequency. Specifically, what is meant by "fast" switching? There are several ways to answer this question; we will only discuss a few of the most important ones here. First, notice from (4.20) and (4.25) that both L_{min} and C_{min} are inversely proportional to f_{sw}. Thus, all else being equal, a higher switching frequency reduces the size of the converter's energy storage components. In a general sense, then, "fast" switching is that which allows the circuit to have reasonably sized components. Next, notice from (4.19) and (4.24) that f_{sw} is also inversely proportional to both ΔI and ΔV. All else being equal, then, faster switching reduces ripple magnitude. Thus, "fast" switching keeps the circuit ripple below a specified threshold. A third way is to consider the Fourier series of v_*, which, according to (4.7) is simply (4.3) multiplied by V_{in}. According to our dc–dc conversion process shown in Figure 4.7, we would like to low-pass filter the square wave v_* to leave only its dc component as V_{out}. Any ripple on the output is unwanted. Together, the inductor and capacitor serve as a low-pass filter to remove it. The higher the switching frequency, the farther it will be in the frequency domain from the dc component and the easier it will be for the filter components to remove it and higher harmonics. Thus, "fast" switching is that which allows the low-pass *LC* output filter to work effectively.

These qualitative descriptions of fast switching are related and essentially describe the same physical circuit behavior; however, we have provided three alternative ways of thinking about the concept to facilitate a deeper understanding. Other considerations that influence the switching frequency include audible noise – humans hear frequencies up to around 20 kHz, so switching frequencies are often chosen higher than this to avoid converter "buzzing," – switch hardware limitations, losses – real switches have small losses each time they switch; the higher the frequency, the higher these losses – and switch gate control hardware limits. These advanced topics lie outside our scope.

Example 4C ***Buck converter design***

Here, we design a buck converter to power a 5 V, 25 W resistive load from a 12 V battery. First, we choose appropriate values for f_{sw}, k, L, and C so that the inductor's peak-to-peak current ripple should be no more than 40% of its average current and the output peak-to-peak voltage ripple should be no more than 4% of the nominal value. Then, for the chosen circuit components, we calculate the expected ΔI and ΔV and compare these to the specifications.

The duty ratio is given by (4.10) and is

$$k = \hat{v}_{out}/V_{in} = 5/12 \tag{4C.1}$$

To avoid the audible range but remain low enough for average switch hardware to accommodate, we will choose $f_{sw} = 30$ kHz. Once the design is complete, we can verify that our components are reasonably sized. If they are not, we can adjust the switching frequency and component design accordingly. The output power is 25 W; thus

$$\hat{i}_{out} = 25\ \text{W}/5\ \text{V} = 5\ \text{A} \tag{4C.2}$$

and from (4.11) $\hat{i}_L = 5$ A. From the specification,

$$\Delta I_{max} = 0.4 \times \hat{i}_L = 2\ \text{A} \tag{4C.3}$$

thus from (4.20)

$$L_{min} = \frac{5 \times (1 - 5/12)}{(30 \times 10^3) \times 2} = 48.61\ \mu\text{H} \tag{4C.4}$$

Let us choose $L = 50\ \mu$H. Equation (4.19) now gives the expected current ripple as

$$\Delta I = \frac{5 \times (1 - 5/12)}{(30 \times 10^3) \times (50 \times 10^{-6})} = 1.94\ \text{A} \tag{4C.5}$$

which is 39% of the nominal current, within the specification. From the specification,

$$\Delta V_{max} = 0.04 \times 5\ \text{V} = 0.2\ \text{V} \tag{4C.6}$$

and from (4.25)

$$C_{min} = \frac{1.94}{8 \times (30 \times 10^3) \times 0.2} = 40.4\ \mu\text{F} \tag{4C.7}$$

Because it is a standard capacitor size, we will choose $C = 47\ \mu$F. Based on this choice, (4.24) gives the expected voltage ripple as

$$\Delta V = \frac{1.94}{8 \times (47 \times 10^{-6}) \times (30 \times 10^3)} = 0.17\ \text{V} \tag{4C.8}$$

which is 3.4% of the nominal output voltage, within the specification. The chosen components are reasonably sized for this type of circuit, and the circuit achieves its design goals. Thus, the switching frequency does not need to be adjusted.

The circuit designed in Example 4C has been simulated, and several waveforms are shown below in Figure 4.14 to illustrate the behavior. Notice that $v_* = q(t)V_{in}$ is as designed, with $k = 5/12$. The period is 33.33 μs, which corresponds to the 30 kHz switching frequency. The inductor current has an average of 5 A and peak-to-peak ripple of approximately 2 A. The input current is equal to the inductor current when the transistor is on and zero otherwise. The output capacitor voltage has an average of 5 V and peak-to-peak ripple of approximately 0.18 V. Notice that the output voltage ripple is out of phase with the inductor current ripple; this is because the capacitor voltage rises when the inductor current is greater than its average, 5 A, and discharges when it is less than its average. The capacitor and output (resistor) current are plotted together to show that the majority of the ripple in i_L goes through the capacitor, while the dc current goes exclusively through the resistor. Finally, the output resistor power is plotted to show the average power is 25 W, as designed, with small peak-to-peak ripple of roughly 2 W, or 8%.

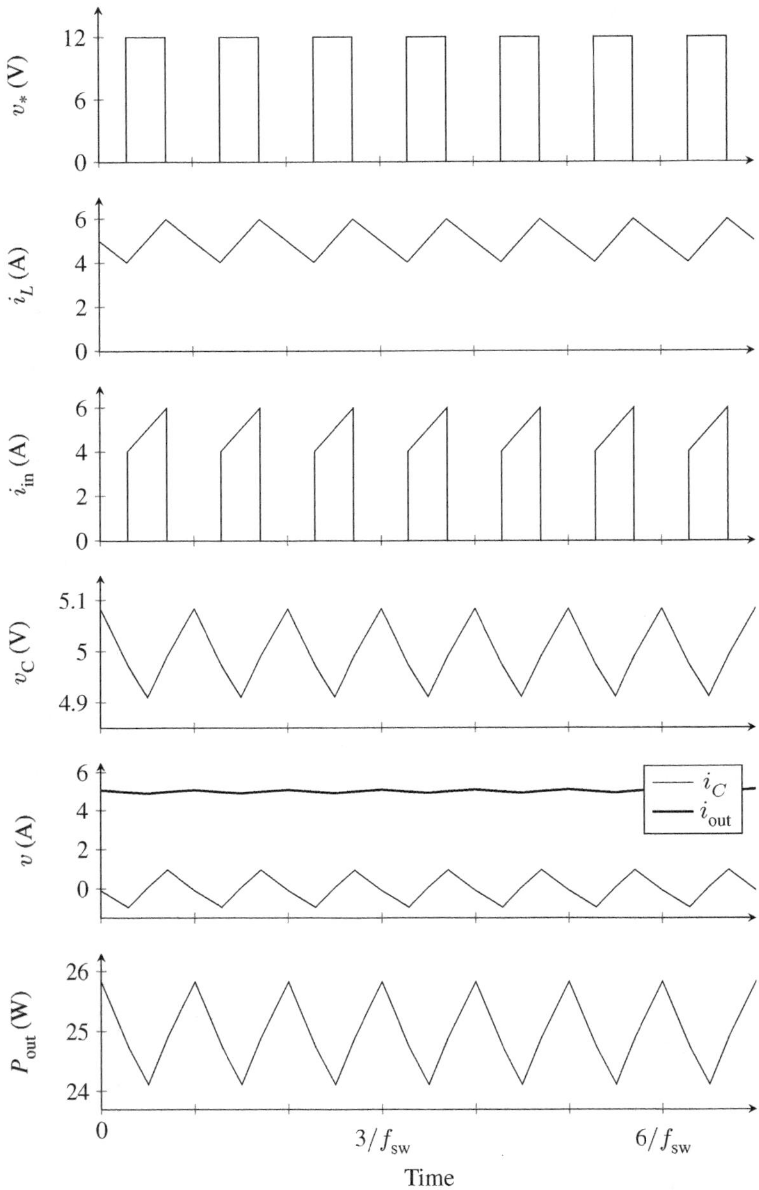

Figure 4.14 Circuit waveforms from the 25 W buck converter designed in Example 4C.

4.3.2 Boost Converter

A companion circuit to the buck converter is the dc–dc *boost converter*, whose basic circuit is shown in Figure 4.15. Its purpose is the inverse of the buck converter: it transfers power from a lower-voltage dc source to a higher-voltage dc load. In contrast to the buck converter, in the boost converter the inductor is always connected in series with the input voltage source. By controlling the transistor with an appropriate switching function, the inductor can be charged and discharged while maintaining an average current. This current is alternately disconnected from and connected to the load through the uncontrolled (diode) switch. The load current is filtered with the output capacitor. With sufficiently "large" L and C, the inductor and capacitor can be approximated as dc current and voltage sources, respectively, to facilitate design and analysis.

Like the buck converter, the boost converter is periodically switched between two circuit configurations. When the transistor is closed, the diode is reverse-biased and turns off. In this circuit configuration, the inductor's voltage is V_{in} and its current is ramped up as energy is transferred from the source to the inductor's magnetic field. The transistor is then opened, which causes the diode to turn on to provide a path for the inductor current to the load. Since $\hat{v}_{\text{out}} > V_{\text{in}}$, the inductor voltage is now negative and its current ramps down as the energy transferred to it from the source is now transferred to the load. The inductor voltage is thus

$$v_L(t) = q(t)V_{\text{in}} + (1 - q(t)) \times (V_{\text{in}} - \hat{v}_{\text{out}}) \tag{4.26}$$

Because the average inductor voltage is zero in the steady state (4.5), we have

$$\begin{aligned}
\hat{v}_L &= \frac{1}{T_{\text{sw}}} \int_0^{T_{\text{sw}}} [q(t)V_{\text{in}} + (1 - q(t)) \times (V_{\text{in}} - \hat{v}_{\text{out}})]dt \\
&= \frac{1}{T_{\text{sw}}} \left[\int_0^{kT_{\text{sw}}} V_{\text{in}} dt + \int_{kT_{\text{sw}}}^{T_{\text{sw}}} (V_{\text{in}} - \hat{v}_{\text{out}})dt \right] \\
&= \frac{1}{T_{\text{sw}}} [V_{\text{in}} \times kT_{\text{sw}} + (V_{\text{in}} - \hat{v}_{\text{out}}) \times (T_{\text{sw}} - kT_{\text{sw}})] = 0
\end{aligned} \tag{4.27}$$

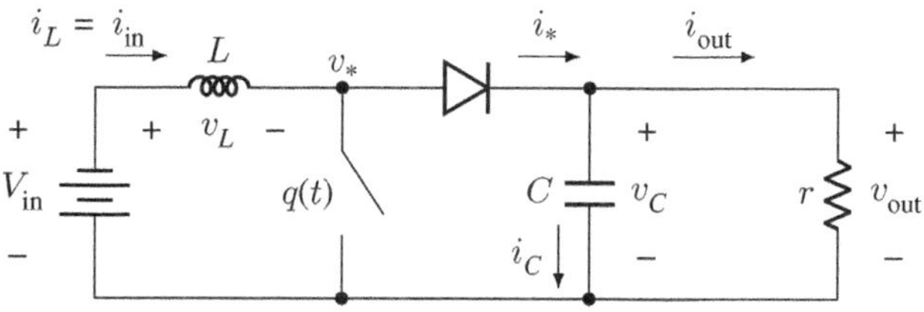

Figure 4.15 Boost converter circuit.

which simplifies as

$$\frac{\hat{v}_{out}}{V_{in}} = \frac{1}{1-k} \tag{4.28}$$

Equation (4.28) describes the steady-state voltage conversion ratio of the boost converter. It is analogous to (4.10) for the buck converter. In the boost converter, the average output voltage is greater than or equal to the input voltage, and the conversion ratio is controlled by k. When $k = 0$, the transistor is never on and the input and average output voltages are equal. At any other value of k, the average output voltage is higher than the input. For example, for $k = 0.5$, $\hat{v}_{out} = 2V_{in}$ and for $k = 0.75$, $\hat{v}_{out} = 4V_{in}$. Note that according to (4.28) $\hat{v}_{out}$ tends toward infinity as k approaches unity. This is because as k increases, more energy is transferred to the inductor while the transistor is in the on-state, and the fraction of the switching period for this energy to be transferred to the load when the transistor is off, $1 - k$, decreases. For this increasing energy to be transferred to the load in a decreasing time, it must be done in short pulses of current at high voltage. In the extreme as k approaches unity, these pulses become infinitesimally small while the voltage becomes infinitely large. Although this description is true in the ideal circuit we have analyzed, in any real converter this is not the case. Series resistance in the inductor will limit the amount of energy that can be transferred to it, bounding the output voltage. This is addressed in Problem 4.7. More generally, the converter's losses will dominate the behavior at higher duty ratios, bounding the output voltage to a finite value. For example, in Figure 4.16, the conversion ratio of an ideal boost converter (4.28) is compared to that of a boost converter with resistive losses in its inductor equal to 1% of the load resistance. As can be seen, in the lossy circuit the boost ratio peaks at 5 with $k = 0.9$, then falls off to zero at $k = 1$.

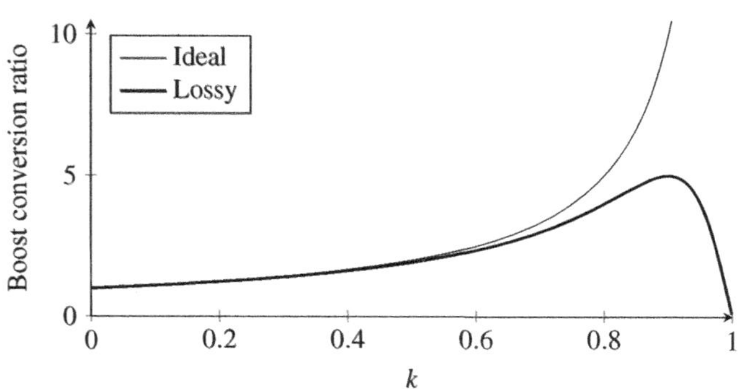

Figure 4.16 Boost converter conversion ratio ($\hat{v}_{out}/V_{in}$) for ideal (lossless) and lossy circuits.

Analysis of the circuit currents and powers in the ideal boost converter reveals that

$$i_{\text{in}} = i_L \tag{4.29}$$

$$\begin{aligned}\hat{i}_{\text{out}} &= (1-k)\hat{i}_L \\ &= (1-k)\hat{i}_{\text{in}}\end{aligned} \tag{4.30}$$

$$\begin{aligned}\hat{P}_{\text{in}} &= V_{\text{in}}\hat{i}_{\text{in}} \\ &= V_{\text{in}}\frac{\hat{i}_{\text{out}}}{1-k} = \frac{V_{\text{in}}}{1-k}\hat{i}_{\text{out}} \\ &= \hat{v}_{\text{out}}\hat{i}_{\text{out}}\end{aligned} \tag{4.31}$$

Assuming v_{out} and i_{out} are approximately constant and equal to $\hat{v}_{\text{out}}$ and $\hat{i}_{\text{out}}$, respectively, we have

$$\hat{P}_{\text{out}} = (\widehat{v_{\text{out}}i_{\text{out}}}) = \hat{v}_{\text{out}}\hat{i}_{\text{out}} \tag{4.32}$$

Thus, $\hat{P}_{\text{in}} = \hat{P}_{\text{out}}$ and power is conserved.

The boost converter power conversion strategy is hinted at by (4.30), and is summarized as follows. In the steady state, the inductor can be approximated as a dc current source; we ignore small ripple current in this discussion. Because $\hat{P}_{\text{in}} = \hat{P}_{\text{out}}$ and $\hat{v}_{\text{out}} > V_{\text{in}}$, this requires that $\hat{i}_{\text{out}} < \hat{i}_{\text{in}}$. This dc current reduction is accomplished by first "chopping" the input current to create the square wave diode current i_*. The dc value of i_* is then obtained by filtering; the whole process being analogous to the voltage reduction strategy used in the buck converter and shown in Figure 4.7. In the boost converter, the circuit reduces dc current the same way the buck converter reduces dc voltage. In both circuits, because power is conserved, when output voltage or current decreases, its corresponding current or voltage must increase.

Circuit waveforms from a 25-W, 5–12-V boost converter with the same inductor and capacitor values as the buck converter in Example 4C are shown in Figure 4.17. The waveforms are arranged so that when compared side-by-side with Figure 4.14, the analogous conversion strategies are evident. For example, the top boost waveform, i_*, is analogous to the top buck waveform, v_*. Each of these is the "chopped" version of the related dc input value.

Choosing appropriate L, C, and f_{sw} in boost converter design requires similar analysis as was done for the buck converter. Inductor current ripple, capacitor voltage ripple, and switch frequency considerations are all used to choose appropriate values. Although we will not explicitly cover this here, you now have the necessary tools to accomplish the task.

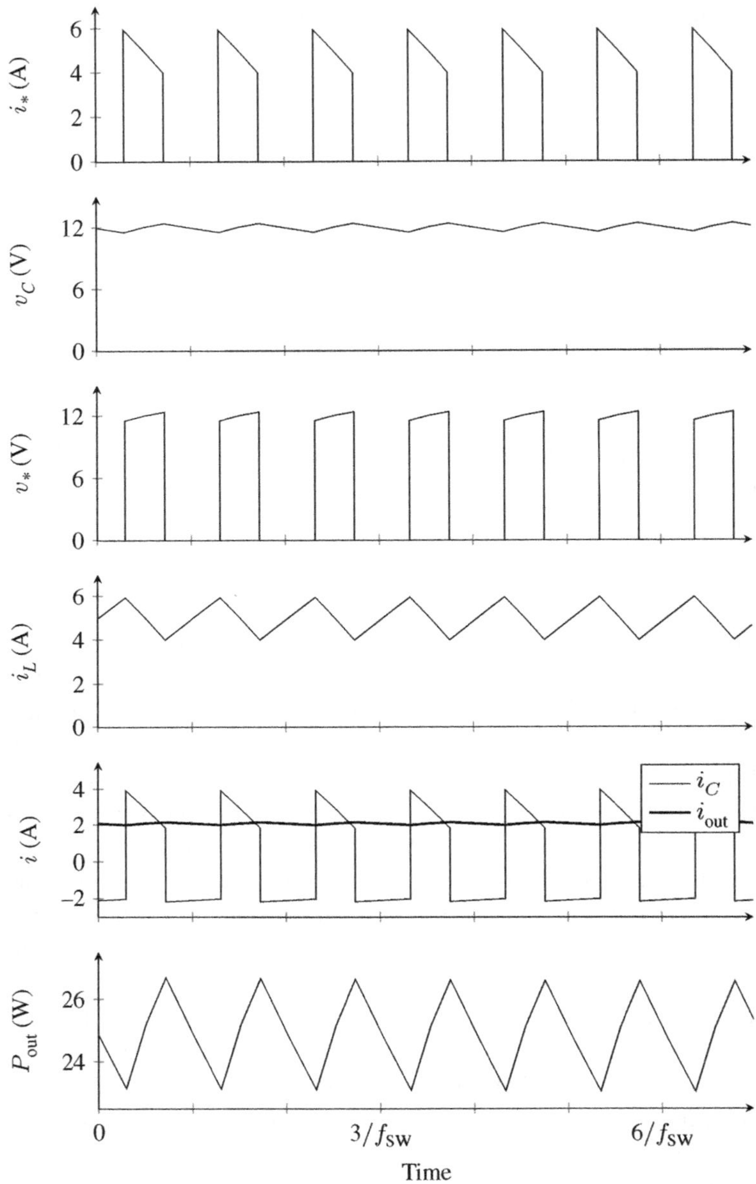

Figure 4.17 Circuit waveforms from ideal 25 W boost converter with $L = 50$ µH, $C = 47$ µF, $f_{sw} = 30$ kHz, $V_{in} = 5$ V, $\hat{v}_{out} = 12$ V, and $k = 7/12$.

4.3.3 Advanced Circuit Topologies

There are numerous dc–dc converter topologies. Buck–boost and boost–buck converters combine elements of the buck and boost converters to facilitate both voltage increase and decrease in a single converter. A flyback converter also allows the voltage to be increased or decreased, but uses a coupled inductor to achieve isolation and to add flexibility in choosing the output voltage polarity. Multi-stage series and parallel configurations are also common when efficiency, ripple, or other design specifications are restrictive. Although the topologies of these circuits are more elaborate than the simple buck and boost circuits introduced here, the basic concepts we have introduced can be applied to their analysis.

SP4.3.1 For a boost converter with $f_{sw} = 40$ kHz, what is T_{sw}? [$T_{sw} = 25$ μs]

SP4.3.2 For the boost converter in SP4.3.1, if the transistor is on for 15 μs of each switch period, what is the duty ratio k? [$k = 0.6$]

SP4.3.3 Assuming the boost converter in SP4.3.1 and SP4.3.2 is ideal, (a) what is the boost voltage conversion ratio? (b) What is the average output voltage, $\hat{v}_{out}$, if $V_{in} = 10$ V? [(a) $\hat{v}_{out}/V_{in} = 2.5$; (b) $\hat{v}_{out} = 25$ V]

SP4.3.4 Assume the boost conversion ratio for a lossy boost converter is $\frac{\hat{v}_{out}}{V_{in}} = \frac{(1-k)r}{(1-k)^2 r + r_L}$ (you are asked to derive this in Problem 4.7). (a) Find the maximum achievable boost ratio. Assume $r = 1\ \Omega$ and $r_L = 0.01\ \Omega$. (b) What is k at this point? (c) Compare your results to the plot in Figure 4.16. [(a) $\hat{v}_{out}/V_{in} = 5$; (b) $k = k_{max} = 0.9$ (c) results should align with the plot]

SP4.3.5 To reverse the process of the boost converter above by converting $V_{in} = 25$ V to $\hat{v}_{out} = 10$ V using a buck converter, what duty ratio k is required? [$k = 0.4$]

4.4 ac–dc Conversion

We now shift our focus to ac–dc conversion circuits, commonly called *rectifiers*. The key concepts are introduced through the analysis of two common circuits: the half- and full-wave rectifier, respectively. The analysis tools introduced in this section are broadly applicable to advanced rectifier circuits.

4.4.1 Half-wave Rectifier

The single-phase *half-wave rectifier* is shown in Figure 4.18. The ac voltage $V_{in}(t) = V_p \sin(\omega_{in} t)$ with radian frequency $\omega_{in} = 2\pi f_{in}$ supplies the load through

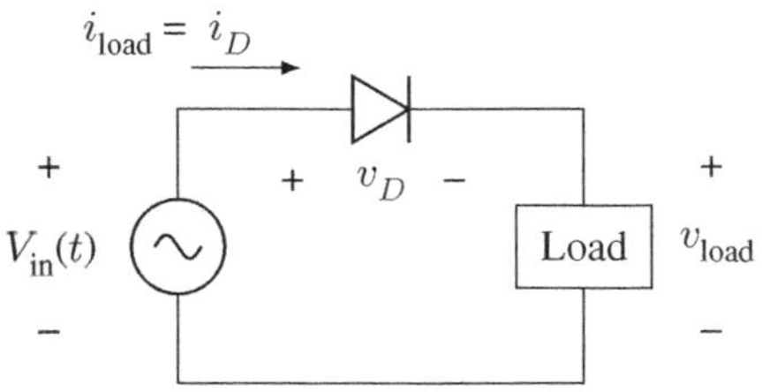

Figure 4.18 Half-wave rectifier circuit.

a diode. The current is i_{load} and the diode and load voltages are v_D and v_{load}, respectively. We will analyze the circuit with loads consisting of various arrangements of passive elements. The diode obeys the rules given in (4.4).

An effective analysis strategy for this circuit is the *trial method* [1], which proceeds as follows: first, we assume a diode state; when the diode is assumed on, we set $v_D = 0$ and then determine the direction of the diode current i_D; if the diode is assumed off, we set $i_D = 0$ and then determine the polarity of v_D. Except possibly at times when the diode changes state, one and only one state will render a circuit configuration that is consistent with the diode rules in (4.4). This is the true diode state, and once it is known the circuit can be solved. Formally, this process must be applied at each instant of time to calculate the circuit's full transient behavior. However, in practice, the behavior can be bracketed into a few discrete time regions. For example, with sinusoidal input the time domain can be divided into three regions: positive, negative, and zero input, respectively. Within each of these regions, the diode state is constant.

To demonstrate the trial method, we assume a resistive load r and solve the circuit in Figure 4.18. When $0 < t < T/2$, $V_{in} > 0$. First, we assume the diode is off and set $i_D = i_{load} = 0$. With this assumption, $v_{load} = i_{load} \times r = 0$ and $v_D = V_{in} - v_{load} = V_{in} > 0$. But, if the diode is truly off, v_D must be negative. Thus, our assumption is inconsistent with (4.4) and the diode must be on. To verify this, we set $v_D = 0$, so that $v_{load} = V_{in} > 0$, and $i_{load} = v_{load}/r > 0$. $i_{load} > 0$ is consistent with the diode being on, which confirms that this is the diode's correct state. Following the same procedure for the interval $T/2 < t < T$ when $V_{in} < 0$, we find that the diode must be off. All that remains is to analyze the instants when $V_{in} = 0$. Unfortunately, the trial method tells us nothing about the diode state in these instants, because both the on and off states lead to inconsistent results. This simply means that the diode is changing state at these instants, transitioning from on to off at $t = T/2$ and from off to on at $t = T$. Because the diode is transitioning at these points, we cannot assign it a definitive state. The full diode trial method analysis procedure is summarized in Table 4.1.

From Table 4.1, the operation of the circuit is clear. When the input voltage is positive the diode is on, the load voltage equals the input voltage, and positive current flows through the load. As the input voltage crosses zero, the diode turns off.

Table 4.1 Summary of diode trial method for half-wave rectifier with resistive load.

t	V_{in}	Diode state assumed	Test	Result	Ok?
$(0, T/2)$	>0	On	$v_D = 0$	$i_D > 0$	Yes
		Off	$i_D = 0$	$v_D > 0$	No
$(T/2, T)$	<0	On	$v_D = 0$	$i_D < 0$	No
		Off	$i_D = 0$	$v_D < 0$	Yes
$T/2$ or T	=0	On	$v_D = 0$	$i_D = 0$	No
		Off	$i_D = 0$	$v_D = 0$	No

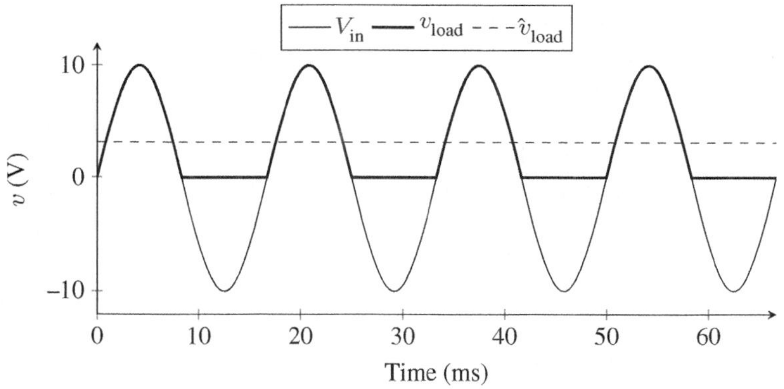

Figure 4.19 Voltage waveforms for a half-wave rectifier with resistive load; $V_p = 10$ V, $r = 1\ \Omega$, and $f_{in} = 60$ Hz.

When the input voltage is negative, the diode is off, no current flows, and the load voltage is zero. The diode switches back on as the voltage crosses zero again one half-period later, and the process continues periodically. Half-wave rectifier input and output voltage waveforms for $f_{in} = 60$ Hz are shown in Figure 4.19. This plot shows why the circuit is named the way it is: half of the input sine wave passes through to the output; the negative half of the wave is blocked.

The average value of this rectified signal is easily calculated as

$$\begin{aligned} \hat{v}_{load} &= \frac{1}{T}\int_0^{T/2} V_p \sin(\omega_{in} t)dt \\ &= \frac{-V_p}{\omega_{in} T}\left[\cos\left(\frac{\omega_{in} T}{2}\right) - \cos(0)\right] \\ &= \frac{V_p}{\pi} \end{aligned} \tag{4.33}$$

In Figure 4.19, $\hat{v}_{\text{load}} = 10/\pi \approx 3.18\,\text{V}$, as plotted. Because it has a non-zero average value, the half-wave rectifier's load voltage is, in fact, a dc signal because it fits our expanded definition of dc from Section 4.2. Because it is periodic, a Fourier series can be constructed for this waveform, and its dc component is V_p/π. The half-wave rectifier converts the ac voltage $V_{\text{in}}(t)$ to the dc voltage v_{load}, with average value $\hat{v}_{\text{load}} = V_p/\pi$. Additional waveforms for circuit current and load power are plotted in Figure 4.20. Current and power only flow to the load when the input voltage is positive and the diode is on.

The analysis is complicated somewhat when the load contains inductors and capacitors, which allow the load to maintain its current and/or voltage independent of the source. This modifies the diode-switching behavior compared to the resistive load case. However, by modifying the diode trial method's resistive load results and using some basic circuit analysis, we can meet the challenge.

First, we examine a parallel *rC* load, which is representative of a bus for dc loads (e.g. a battery charging circuit) or a dc link feeding additional power electronic circuitry. If the capacitance is "large" (i.e. $\tau = rC \gg T$), the capacitor will charge up when the diode is on, but will maintain its charge and slowly discharge approximately linearly through the resistor when the diode is off. If C is "small" the capacitor will discharge quickly when the diode is off and the load behavior will not differ significantly from the resistive case. With large C, the diode will be reverse-biased and remain off until the input voltage is greater than the decaying capacitor voltage, at which point the diode will turn on again. In essence, with respect to the resistive load circuit, the introduction of the capacitor delays the turn-on time of the diode until the input voltage is greater than the capacitor

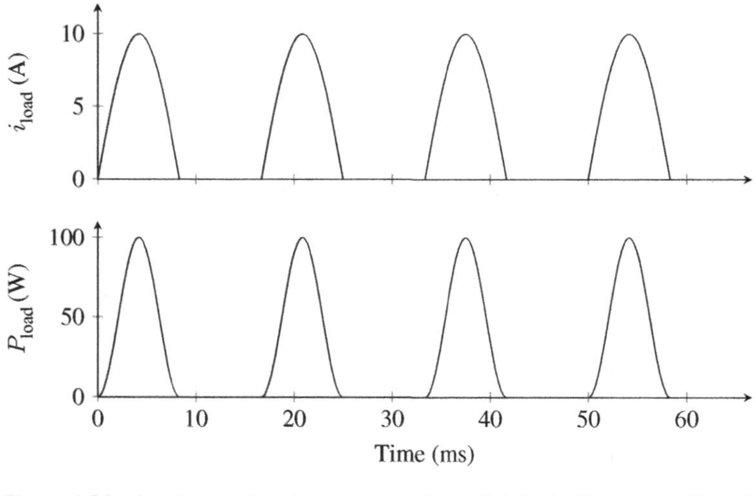

Figure 4.20 Load current and power waveforms for the half-wave rectifier with resistive load.

voltage and accelerates the turn-off time to when the diode current crosses zero (when the capacitor and resistor currents balance exactly), which occurs while the input voltage is still positive. Because the source current leads the source voltage in this capacitive circuit, the diode current always crosses zero before the source voltage.

Calculating the diode turn-on and turn-off times t_{on} and t_{off}, respectively, requires some calculus, which we will omit. However, it suffices to understand conceptually the differences between the resistive and rC load circuits. Waveforms for a half-wave rectifier with a parallel rC load with $r = 1\,\Omega$ and $C = 47$ mF ($\tau = 47$ ms) and source parameters $V_p = 10$ V and $f_{in} = 60$ Hz ($T \approx 16.67$ ms) are given in Figures 4.21 and 4.22. For this circuit, $\tau/T \approx 2.82$. The capacitor's effect is clear: compared to a resistive load, $\hat{v}_{load}$ is much higher and the ripple about the average is much lower. The capacitor acts as both a low-pass filter and a charge storage device increasing and smoothing the resistor voltage.

In Figure 4.22, it is shown that the parallel rC load draws current in short, large bursts when the diode is on. Compared to Figure 4.20, the current and power pulses are over 10 times larger than with a resistive load. Because these large pulses are common in diode capacitor circuits, filters are often used to improve power quality, as is described later in this section.

Next, let us consider a series rL load. This might represent, for example, the field winding of either a synchronous or dc machine. If the inductance is "small"[1] (i.e. $\tau = L/r \ll T$), the source current will lag the voltage by a small angle much less than $\pi/2$. The current zero-crossing, and the corresponding diode turn-off,

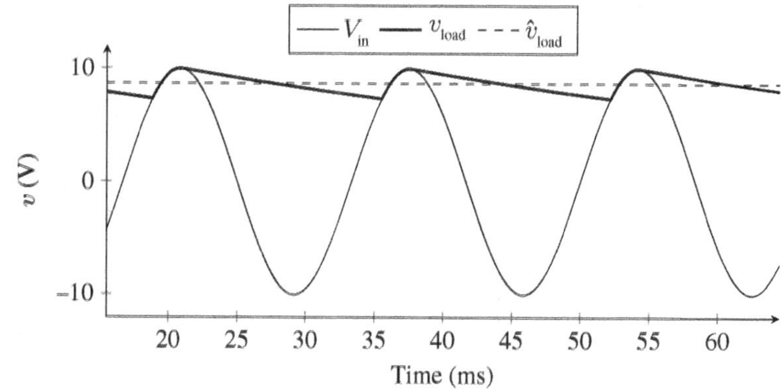

Figure 4.21 Steady-state voltage waveforms from a half-wave rectifier with parallel rC load; $V_p = 10$ V, $r = 1\,\Omega$, $C = 47$ mF, and $f_{in} = 60$ Hz.

1 We only consider the small inductance case, because with large L the inductor current will not go to zero during the half-cycle when the input voltage is negative; thus, the diode will never turn off, and rectification will not occur.

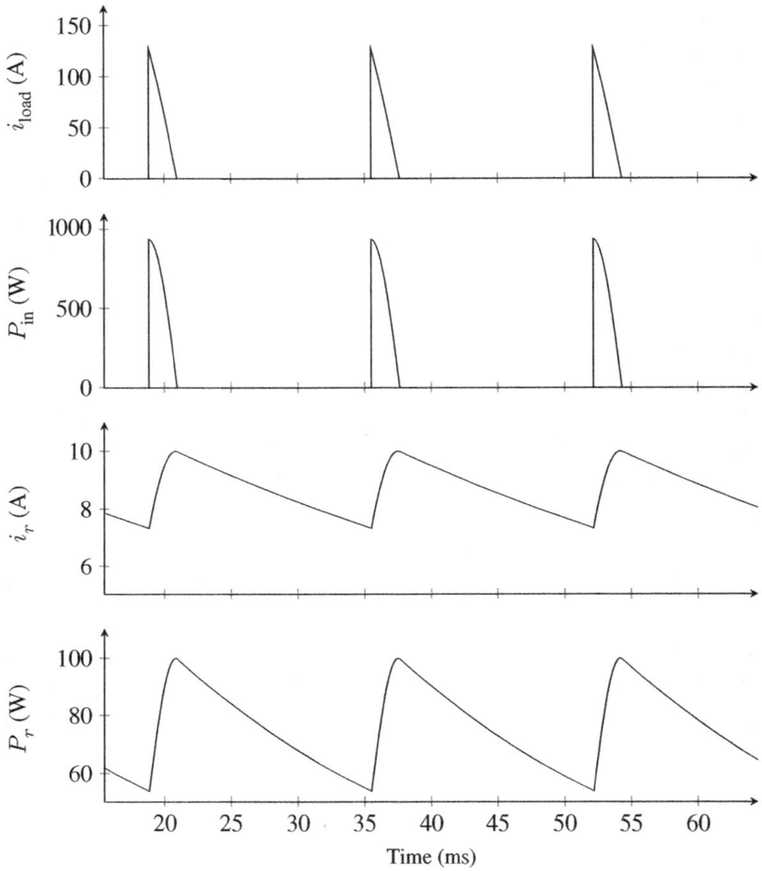

Figure 4.22 Source current and power, and load (resistor) current, and power waveforms for the half-wave rectifier with parallel *rC* load.

will lag the voltage zero crossing. For a short time during each cycle, the source voltage will be negative while the current is positive: energy will flow from the load to the source. Once the diode turns off, the load conditions are identical to the off-state conditions of the resistive load circuit; thus, the diode will turn on at the same times. However, the turn-off will be delayed, which will affect the average rectified load voltage.

As with the *rC* load, explicitly calculating the diode turn-off times requires non-trivial calculations that will not be shown here. However, a firm conceptual understanding of the impact of the inductor's addition is achieved by studying Figures 4.23 and 4.24 below, which plot waveforms from a half-wave rectifier with series *rL* load, the same parameters as the resistive circuit and $L = 3\,\text{mH}$ ($\tau = 3\,\text{ms}$). For this circuit, $\tau/T \approx 0.18$. Comparing Figure 4.23 to its

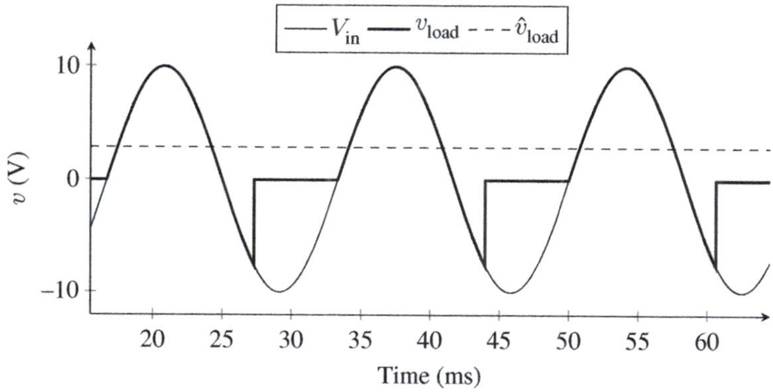

Figure 4.23 Steady-state voltage waveforms for a half-wave rectifier with series rL load; $V_p = 10$ V, $r = 1\,\Omega$, $L = 3$ mH, and $f_{in} = 60$ Hz.

resistive (Figure 4.19) and parallel rC (Figure 4.21) counterparts, and comparing Figure 4.24 to Figures 4.20 and 4.22 shows how the addition of energy storage elements impacts the performance of the half-wave rectifier.

Example 4D ***Average power in a half-wave rectifier***

We now calculate the average power delivered by the half-wave rectifier to a resistive load r.

The relevant circuit waveforms are shown in Figures 4.19 and 4.20. The load power, P_{load}, is expressed as

$$P_{load} = \frac{v_{load}^2}{r} \tag{4D.1}$$

The average power delivered to the load is

$$\hat{P}_{load} = \frac{\widehat{(v_{load}^2)}}{r} \tag{4D.2}$$

which is calculated as

$$\begin{aligned} \hat{P}_{load} &= \frac{1}{r}\frac{1}{T_{in}}\int_0^{T_{in}/2} V_p^2 \sin^2(\omega_{in}t)dt \\ &= \frac{1}{r}\frac{V_p^2}{T_{in}}\int_0^{T_{in}/2} \sin^2(\omega_{in}t)dt \end{aligned} \tag{4D.3}$$

The identity

$$\sin^2(\omega_{in}t) = \frac{1}{2}[1 - \cos(2\omega_{in}t)] \tag{4D.4}$$

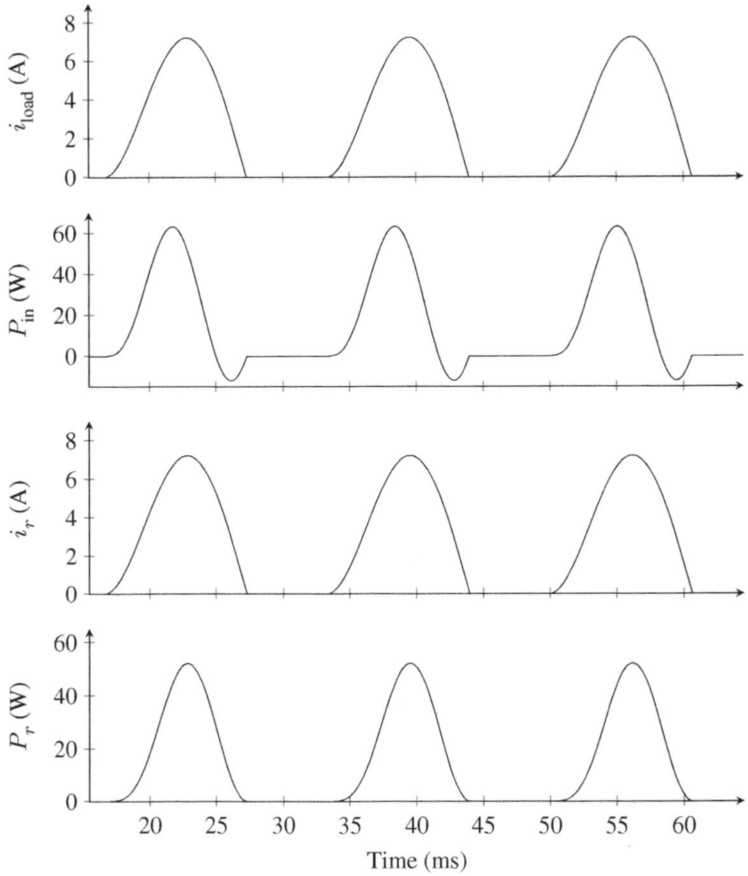

Figure 4.24 Source current and power, and load (resistor) current and power waveforms for the half-wave rectifier with series *rL* load.

is used to integrate (4D.3), giving

$$\widehat{P}_{\text{load}} = \frac{V_p^2}{4r} \tag{4D.5}$$

4.4.2 Full-wave Rectifier

From Figure 4.19, it is apparent that the average load voltage obtained from the half-wave rectifier can be doubled by flipping the clipped negative humps, making them positive, and applying them to the load as well. In other words, we would like to create $v_{\text{load}} = |V_{\text{in}}|$. This is accomplished by adding three diodes and forming a *full-wave rectifier* (also called a *full-bridge rectifier*). The circuit is shown in Figure 4.25.

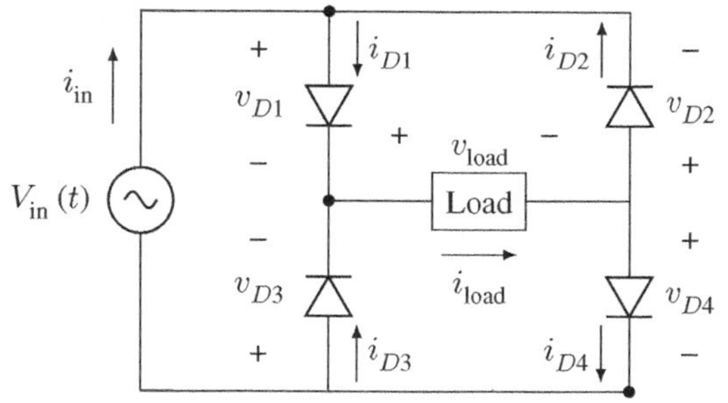

Figure 4.25 Full-wave rectifier circuit.

The full-wave rectifier can be analyzed using the trial method: we first assume a resistive load and analyze all the diode state combinations, both for positive, negative, and zero input voltage. Because there are now four diodes, there are $2^4 = 16$ switch combinations for each of the three input voltage regions, leaving 48 combinations to analyze. Although this seems daunting at first, many of the combinations can be eliminated quickly.

Like the half-wave rectifier, when $V_{in} = 0$ none of the 16 diode states is self-consistent. These are the instants when the diodes change state. Of the remaining 32 configurations, several are quickly eliminated due to Kirchhoff's voltage law (KVL) and KCL violations. Any configuration with only one diode on violates KCL, because there is not a closed loop for current to flow. This fact eliminates eight configurations (four with $V_{in} > 0$ and four with $V_{in} < 0$). Likewise, any configuration with diodes 1 and 3 on simultaneously and/or diodes 2 and 4 on simultaneously violates KVL, because these configurations short-circuit the input source and require negative current to flow through one of the diodes. This eliminates 14 more configurations. In the two cases with all four diodes off, no load current can flow, which means that $v_{D1} = -v_{D2}$ and $v_{D3} = -v_{D4}$. This is inconsistent with all off diode voltages being negative, eliminating these two cases. There remain only eight configurations to examine.

For simplicity, the collective diode state assumed is represented by a four-digit binary number, where 1 = on and 0 = off. For example, the diode state with diodes 2 and 3 on and diodes 1 and 4 off is represented as 0110. After applying the trial method to the remaining eight cases, only two cases – when $V_{in} > 0$ and diodes 1 and 4 are on, and when $V_{in} < 0$ and diodes 2 and 3 are on – are consistent. The two key circuit configurations resulting from the trial method for the full-wave rectifier are given in Table 4.2. The full analysis is left to the reader as an end-of-chapter problem.

The full-wave rectifier operation is now evident. As time increases from 0 to $T/2$ the circuit is in diode state 1001, the positive half-cycle of the sinusoid is applied across the load and current flows through diodes 1 and 3 and the load. The diodes switch state as the voltage crosses zero at $t = T/2$. The input voltage sinusoid is negative from $T/2$ to T, and $-V_{in}$ (which is positive) is applied across the load.

Table 4.2 Summary of defining states in the diode trial method for a full-wave rectifier with resistive load.

			Test result				
t	V_{in}	**Diode state**	***D1***	***D2***	***D3***	***D4***	**Ok?**
$\left(0, \frac{T}{2}\right)$	>0	1001	$i_{D1} > 0$	$v_{D2} < 0$	$v_{D3} < 0$	$i_{D4} > 0$	Yes
$\left(\frac{T}{2}, T\right)$	<0	0110	$v_{D1} < 0$	$i_{D2} > 0$	$i_{D3} > 0$	$v_{D4} < 0$	Yes
$\frac{T}{2}$ or T	=0	Indeterminate diode transition state					

Current flows through diodes 3 and 4 and the load. As the input voltage crosses zero again, the diodes switch back to their original state, and the cycle continues.

The diode trial method thus far presented may seem tedious; however, once we apply it to a few circuits, we develop an intuition for how diode circuits work. The tedium of checking every switch state can then be bypassed as we quickly identify the acceptable configurations. Many introductory texts bypass any discussion of this rigorous method in favor of simply stating how the full-wave rectifier operates; however, a thorough understanding of the trial method is very useful when we encounter a complicated circuit containing diodes. Thus, time invested now will pay dividends down the road.

Input, load, and average voltage waveforms for the full-wave rectifier with resistive load are shown in Figure 4.26. Performing the averaging calculation in (4.33) on v_{load} gives

$$\hat{v}_{load} = \frac{2V_p}{\pi} \tag{4.34}$$

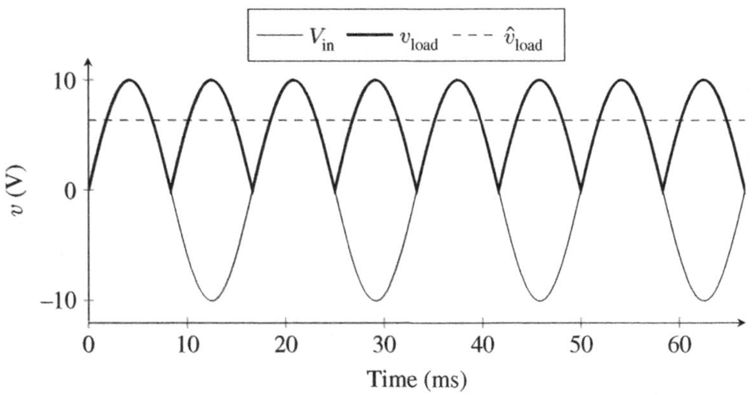

Figure 4.26 Voltage waveforms from a full-wave rectifier with resistive load; $V_p = 10$ V, $r = 1\ \Omega$, and $f_{in} = 60$ Hz.

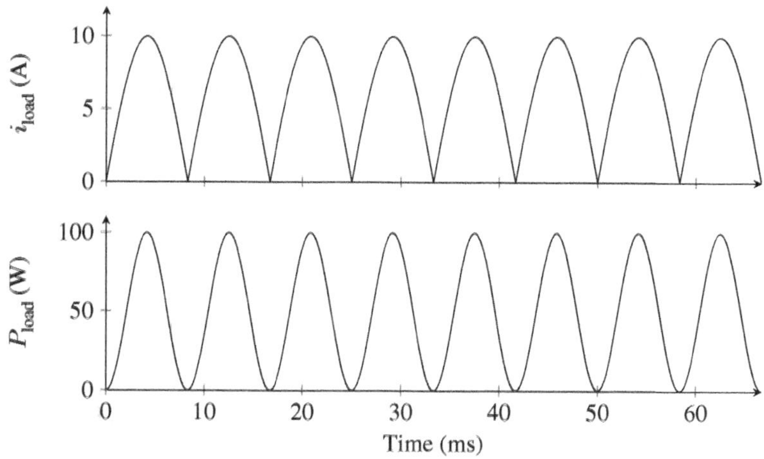

Figure 4.27 Load current and power waveforms for the full-wave rectifier with resistive load.

which is twice the average value of the half-wave rectifier output. Figure 4.27 contains load current and power waveforms. Comparing Figures 4.26 and 4.27 to Figures 4.19 and 4.20, respectively, reveals that the full-wave rectifier output waveforms have double the frequency of the half-wave rectifier. The Fourier series representation of the full-wave rectified waveform shown in Figure 4.26, labeled $v_{rec}(t)$, is given by

$$v_{rec}(t) = \frac{2V_p}{\pi} + \frac{4V_p}{\pi}\sum_{n=1}^{\infty}\frac{\cos(n\pi)}{1-4n^2}\cos[n(2\omega_{in}t - 2\theta(0))] \tag{4.35}$$

where $V_p = 10$ V, $\omega_{in} = 2\pi \times 60$ rad/s, and $\theta(0) = \pi/2$. Notice that the dc component of the Fourier series matches the average value in (4.34), and the series' fundamental frequency is twice the frequency of V_{in}. When compared to the half-wave rectifier output, the double-frequency harmonics of the full-wave rectifier are relatively easier to filter.

The effect of adding parallel capacitance to the full-wave rectifier output is similar to its impact on the half-wave rectifier. However, since the full-wave rectified voltage has twice the frequency of the half-wave, the filtered voltage has a higher average and reduced ripple magnitude. With a capacitor in parallel with the load, the circuit is often called a *classical rectifier*, because this is a classical (i.e. common) charging circuit for batteries. Waveforms for the full-wave rectifier shown in Figure 4.25 with 47 mF parallel load capacitance are shown in Figures 4.28 and 4.29. Comparing these figures to the half-wave rectifier waveforms in Figures 4.21 and 4.22, respectively, clearly shows the double frequency and reduced load ripple.

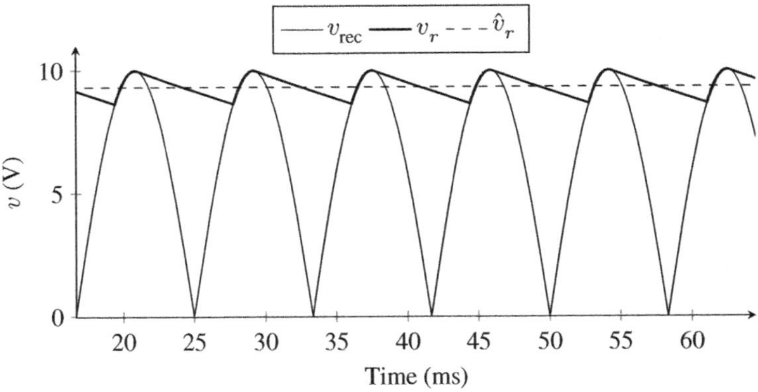

Figure 4.28 Voltage waveforms from a full-wave rectifier with parallel rC load; $V_p = 10\,\text{V}$, $r = 1\,\Omega$, $C = 47\,\text{mF}$, and $f_{in} = 60\,\text{Hz}$.

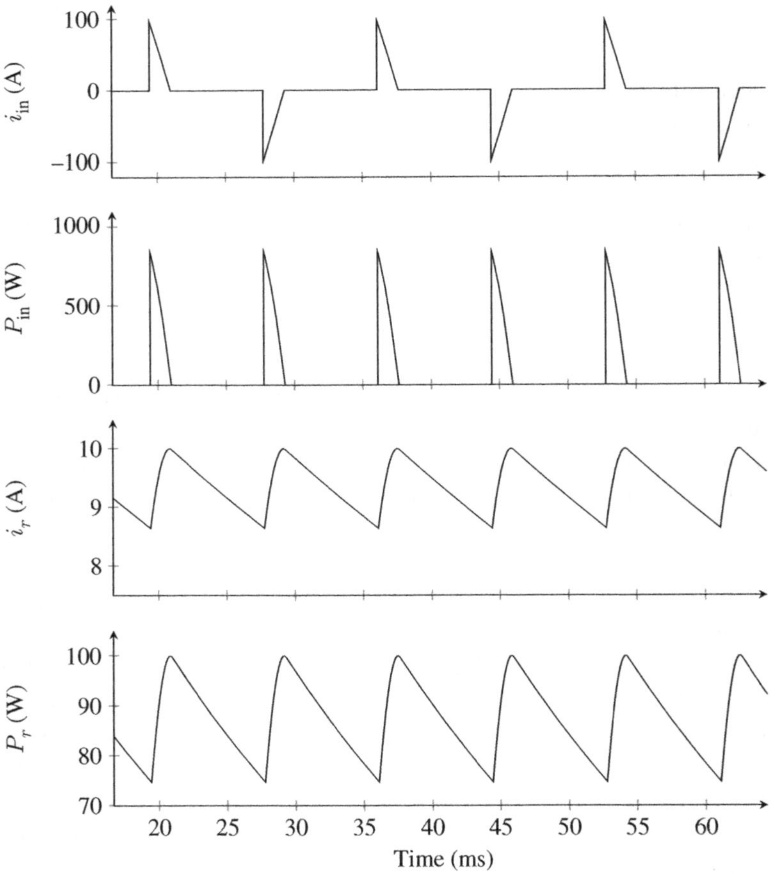

Figure 4.29 Source current and power, and load (resistor) current and power waveforms for the full-wave rectifier with parallel rC load.

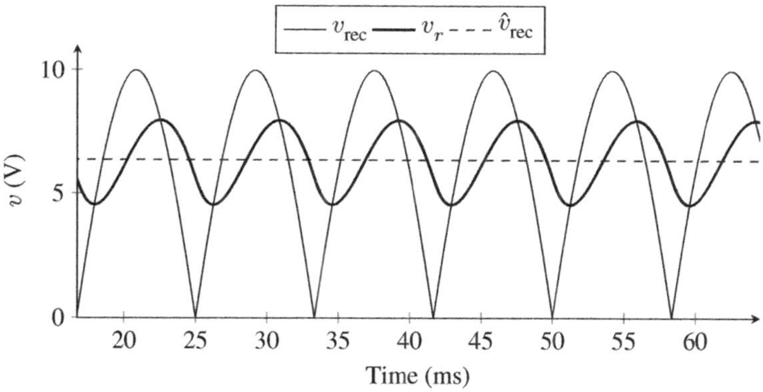

Figure 4.30 Steady-state voltage waveforms from a full-wave rectifier with series rL load; $V_p = 10$ V, $r = 1\ \Omega$, $L = 3$ mH, and $f_{in} = 60$ Hz.

Adding series inductance to the output has effects similar to the half-wave rectifier, with one key difference. Because there are four switches instead of one, the full-wave rectifier always has an available current path, so there are no diode-switching delays and the current need not decay to zero for the diodes to switch states. This is illustrated in Figures 4.30 and 4.31. Comparing these to Figures 4.23 and 4.24, respectively, shows that the rectified voltage is not altered by diode-switching delays, the input power is always non-negative, and the load current remains positive. The inductor acts as a low-pass filter that smooths the resistor voltage.

The full-wave rectifier contains four diodes and provides a path for the load current in each of its allowed diode switch states. This facilitates the implementation of more sophisticated load filtering to achieve the desired load power quality. To see this, we consider the LC low-pass filter shown in Figure 4.32. The input to the filter is $v_{rec}(t) = |V_{in}(t)|$, which consists of a dc component equal to $2V_p/\pi$ and harmonics at multiples of $2\omega_{in}$. To achieve true ac-to-dc conversion, all the harmonics above dc should be filtered out. A straightforward steady-state analysis of the circuit (this is left to the reader in SP4.4.2) reveals that the filter transfer function $G_{filt}(\omega)$ is

$$G_{filt}(\omega) = \frac{v_r(\omega)}{v_{rec}(\omega)} = \frac{r}{j\omega L + r(1 - \omega^2 LC)} \tag{4.36}$$

The magnitude of the complex G_{filt}, notated as $||G_{filt}||$, is plotted in Figure 4.33. Also plotted are the Fourier series harmonic coefficients, normalized to the dc component $\frac{2V_p}{\pi}$. With $L = 3$ mH and $C = 47$ mF, the harmonics of $v_{rec}(t)$ are well to the right of the resonance frequency, $f_{res} = \frac{1}{2\pi\sqrt{LC}} \approx 13.4$ Hz, in the filter's

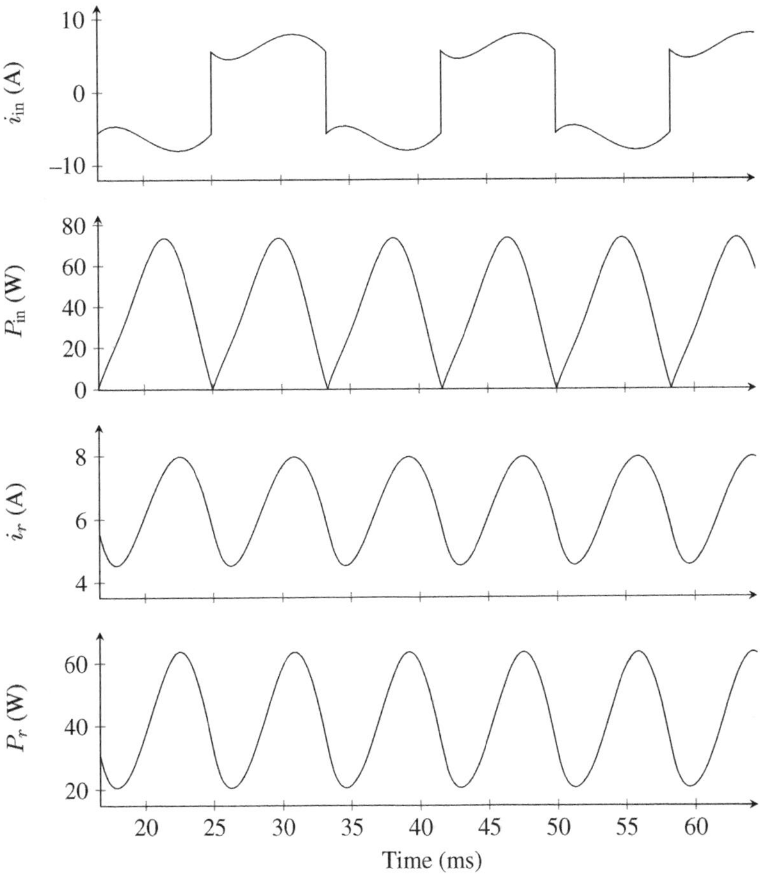

Figure 4.31 Source current and power, and load (resistor) current and power waveforms for the full-wave rectifier with series *rL* load.

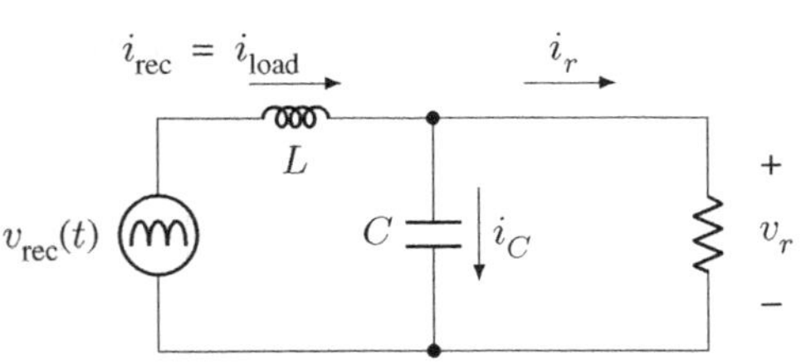

Figure 4.32 Full-wave rectifier with *LC* output filter and resistive load.

capacitive region. The harmonics are thus filtered out effectively, leaving the dc component to dominate the filter output.

Voltage waveforms for the filtered full-wave rectifier are shown in Figure 4.34. The load voltage v_r closely matches $\hat{v}_{rec}$, highlighting the *LC* low-pass filter effects and the circuit's ac–dc converter action. In general, *L* and *C* should be chosen so

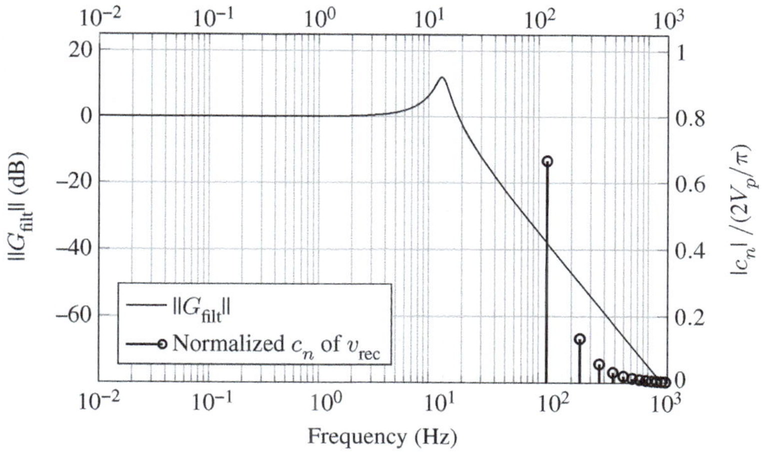

Figure 4.33 $\|G_{\text{filt}}\|$ and normalized Fourier series harmonic coefficients (c_n) of v_{rec} versus frequency for a full-wave rectifier with *LC*-filtered resistive load and $V_p = 10$ V, $L = 3$ mH, $C = 47$ mF, and $r = 1\ \Omega$.

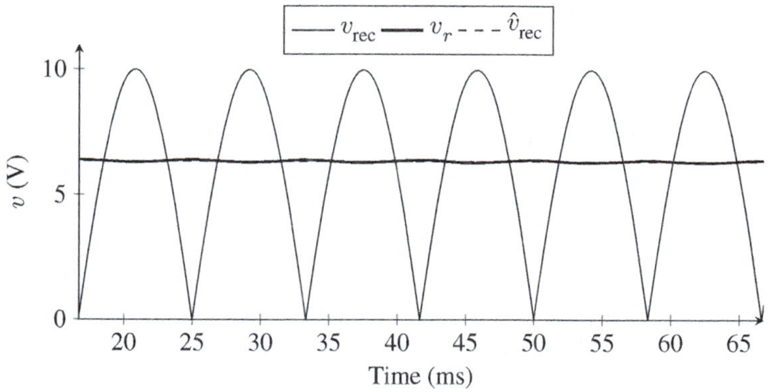

Figure 4.34 Steady-state voltage waveforms from a full-wave rectifier with *LC*-filtered resistive load; $V_p = 10$ V, $r = 1\ \Omega$, $L = 3$ mH, $C = 47$ mF, and $f_{\text{in}} = 60$ Hz.

that the harmonics of the rectified waveform are located well into the capacitive region of the filter to the right of the resonance frequency; otherwise, the filter's attenuation will be minimal. This again illustrates the essence of power electronics: power conversion is achieved by chopping an input signal using switches and filtering the chopped signal before applying it to a load. In the buck converter, the input voltage was chopped into a square wave and filtered; in the boost converter, the input current was chopped into a square wave and filtered; in the full- and half-wave rectifiers, the input voltage sinusoid is chopped into half-waves with diodes, and the waves are filtered and applied to a dc load.

SP4.4.1 (a) Calculate the average power delivered by the full-wave rectifier to a resistive load. The input voltage is $V_{in}(t) = V_p \sin(\omega_{in} t)$, where $T_{in} = 2\pi/\omega_{in}$ is the input period. Relevant circuit waveforms are shown in Figures 4.26 and 4.27. (b) Compare your answer in (a) to the result in Example 4D. [(a) $\hat{P}_{load} = V_p^2/2r$; (b) the average power in the full-wave rectifier is twice that in the half-wave rectifier]

SP4.4.2 Derive (4.36) using Figure 4.32.

SP4.4.3 Why does the Fourier series of v_{rec} in (4.35) have a fundamental frequency of $2\omega_{in}$? [$|V_{in}|$ has twice the frequency of V_{in} (the absolute value of a sinusoid has twice the frequency of the sinusoid itself)]

4.5 dc–ac Conversion

We conclude by introducing dc–ac conversion circuits, called *inverters*. Here, we analyze the single-phase ac-to-dc inverter; the basic three-phase inverter is covered in Chapter 5 as part of a brushless dc drive.

4.5.1 Single-phase Inverter

A single-phase dc–ac inverter is shown in Figure 4.35. This circuit converts energy from the dc source, V_{in}, to the ac load using a bridge of four controlled transistor switches, notated with the capital letter S. There are numerous switch control methods, but we will only cover one here and leave the others to an advanced course.

It is a straightforward process to identify the allowed and useful switch states. First, switches 1 and 3 and 2 and 4, respectively, can never be on simultaneously, lest they short the source. Thus, their respective switching functions obey $q_1(t) = 1 - q_3(t)$ and $q_2(t) = 1 - q_4(t)$. Second, if switches 1 and 2 or 3 and 4, respectively, are on simultaneously, the load is essentially short-circuited and

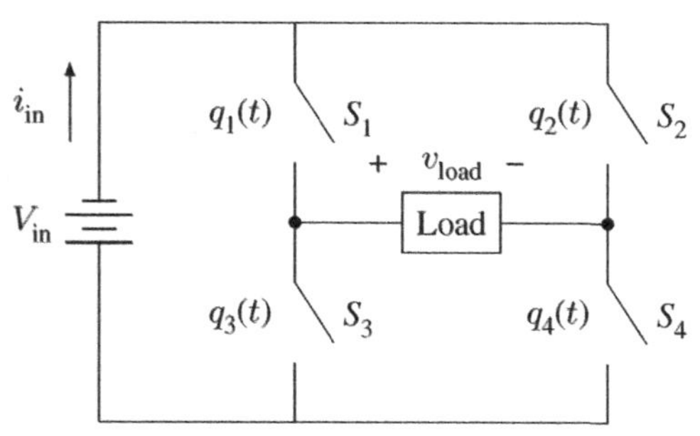

Figure 4.35 Single-phase dc–ac inverter circuit.

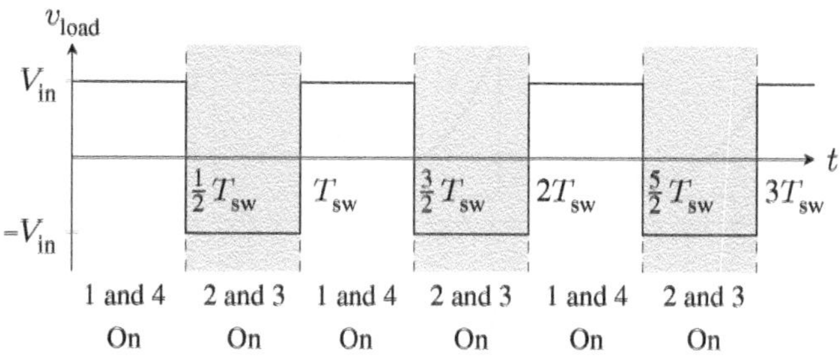

Figure 4.36 Single-phase inverter voltage.

no power will be transferred. Further, if only one (or no) switches are on, there will not be a continuous path for current to flow through the load. Thus, the only allowable switch states that transfer power from the source to the load are $(q_1, q_2, q_3, q_4) = (1, 0, 0, 1)$ and $(0, 1, 1, 0)$. If the circuit is switched between these two states alternately with equal time in each state, the square wave voltage shown in Figure 4.36 will be applied across the load.

Although not a sinusoid, the inverter voltage square wave shown in Figure 4.36 is an ac waveform. Our earlier Fourier analysis showed that this periodic waveform can be expressed as a sum of sinusoids: the fundamental and its harmonics. The series in (4.3) can be easily modified to represent the inverter square wave (see SP4.5.1). Ideally, the harmonics can be removed through appropriate filtering, leaving only a sinusoid with the fundamental frequency of the square wave. In practice, such filtering is difficult to achieve because the harmonics are relatively close to the fundamental in the frequency domain. Advanced courses introduce switching methods based on *pulse width modulation* (PWM), which are used to separate the unwanted harmonics from the fundamental and make it easier to filter them. We will limit our discussion to the simple square wave switching scheme and analyze the circuit with both inductive and resonant *LC* load filters applied.

Voltage, current, and power waveforms are shown in Figure 4.37 for the single-phase inverter with series *rL* load and $V_{in} = 10\,\text{V}$, $r = 1\ \Omega$, $L = 3\,\text{mH}$, and $f_{in} = 60\,\text{Hz}$. The load time constant $\tau = L/R = 3\,\text{ms}$, which is about one-sixth of the switching period T_{sw} (16.7 ms), and the response is as expected: a sequence of *rL* charges and decays. The load current is slightly more sinusoidal than the square wave current that would result without the inductive filtering, but we are still far from a sinusoid. Luckily, we can do better.

A load waveform closer to a sinusoid can be formed by adding a capacitor in series with the series *rL* load and choosing *L* and *C* to resonate at the inverter switch frequency. This creates a series *resonant filter*. For example, by adding series capacitance *C* such that $\frac{1}{2\pi\sqrt{LC}} = 60\,\text{Hz}$, the waveforms in Figure 4.38 result. The

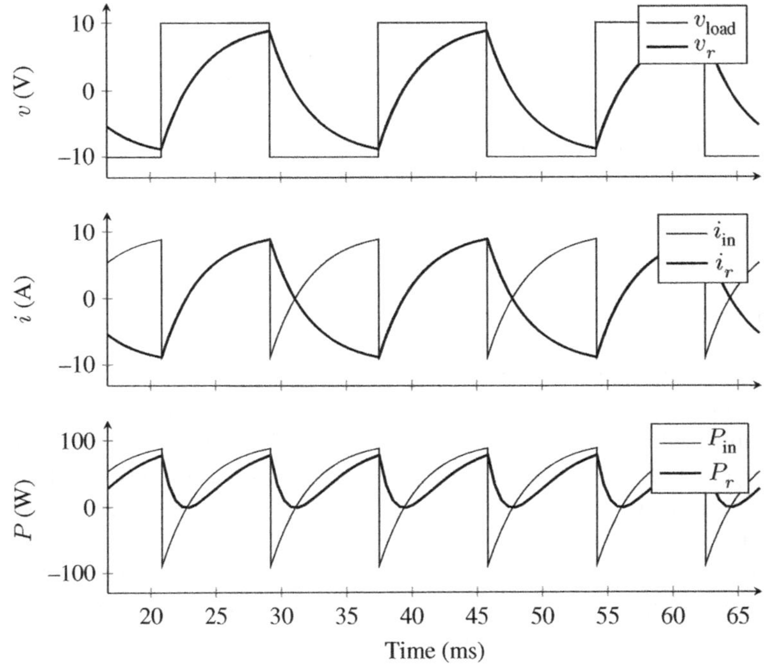

Figure 4.37 Steady-state voltage, current, and power waveforms from a single-phase inverter with series rL load; $V_{in} = 10\,\text{V}$, $r = 1\,\Omega$, $L = 3\,\text{mH}$, and $f_{in} = 60\,\text{Hz}$.

shapes of these load voltage and current waveforms are more sinusoidal than their counterparts in Figure 4.37, but they are not pure sinusoids. However, often these are acceptable ac waveforms. For example, many low-cost consumer single-phase inverters create similar waveforms.

Example 4E ***Power calculation for single-phase inverter***

It is instructive to calculate the average power delivered to a resistive load by the single-phase inverter in Figure 4.35. The load current is simply the square wave in Figure 4.36 scaled by the load resistance. The power delivered to the load is calculated as

$$P_{load} = i_{load}^2 \times r \tag{4E.1}$$

The average power is

$$\widehat{P}_{load} = \widehat{(i_{load}^2 \times r)} = \widehat{(i_{load}^2)} \times r \tag{4E.2}$$

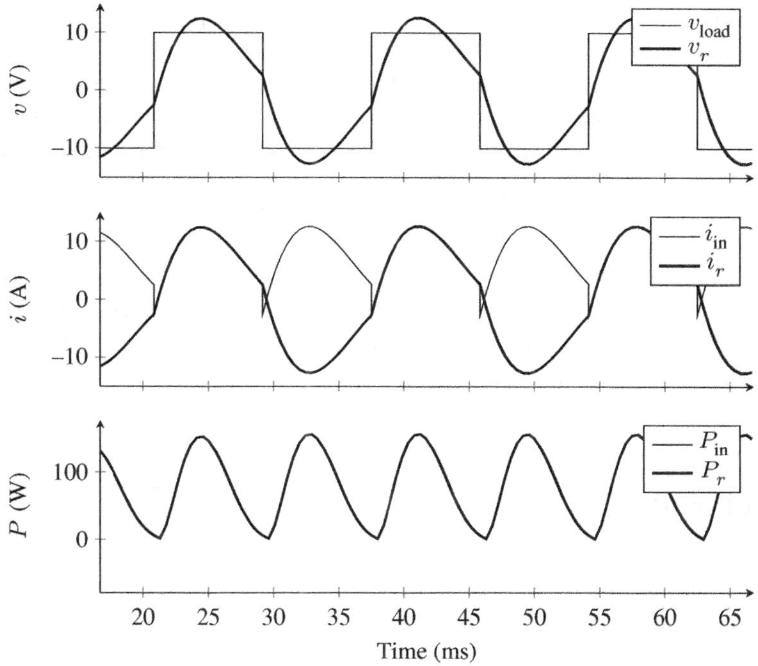

Figure 4.38 Steady-state voltage, current, and power waveforms for the single-phase inverter with series resonant *rLC* load; $V_{\text{in}} = 10\,\text{V}$, $r = 1\,\Omega$, $L = 3\,\text{mH}$, $C = 2.345\,\text{mF}$, and $f_{\text{in}} = 60\,\text{Hz}$.

Notice that this is *not* equivalent to $(\widehat{i}_{\text{load}})^2$, which is zero. We can calculate the load current's mean square, denoted $(\cdot)_{ms}$, as

$$\begin{aligned}\widehat{P}_{\text{load}} &= r \times (i_{\text{load}})_{ms} = \frac{r}{T_{\text{sw}}} \int_0^{T_{\text{sw}}} i^2_{\text{load}} dt \\ &= \frac{r}{T_{\text{sw}}} \int_0^{T_{\text{sw}}} \left(\frac{V_{\text{in}}}{r}\right)^2 dt \\ &= \frac{V^2_{\text{in}}}{r}\end{aligned} \tag{4E.3}$$

We conclude this section by noting that the single-phase inverter can be expanded in a straightforward manner to a three-phase inverter, a device which is common in electric drives. This device, often referred to as a "six-step" inverter, is covered in Chapter 5, where it is used to drive the permanent-magnet ac machine as a brushless dc machine. The concepts covered here with respect to the single-phase inverter apply to that device.

SP4.5.1 Modify (4.3) to find the Fourier series of the single-phase inverter voltage waveform shown in Figure 4.36. First, show that $\theta(0) = \pi/2$, and find k. Plug these values into (4.3), scale $q(t)$ to achieve the appropriate magnitude, and shift the result to remove the scaled dc offset. Finally, use trigonometric identities and properties of trigonometric functions to simplify. $\left[v_{\text{load}}(t) = \frac{4V_{\text{in}}}{\pi} \sum_{n=1,3,5,\ldots}^{\infty} \frac{1}{n} \sin\left(\frac{2\pi nt}{T_{\text{sw}}}\right)\right]$

SP4.5.2 Qualitatively describe any changes to the waveforms in Figure 4.37 if L is reduced to 0.3 mH (90% reduction) [v_r and i_r would look much more like square waves; i_{in} and P_{in} would have much faster rise times; P_r would have less ripple (would look closer to a constant value)]

4.6 Problems

4.1 Show that the efficiency, defined as $\eta = P_{\text{out}}/P_{\text{in}}$, for the resistive voltage divider shown in Figure 4.39, can be expressed as

$$\eta = \frac{r_2^2 r_{\text{out}}}{(r_2 + r_{\text{out}})(r_1 r_2 + r_1 r_{\text{out}} + r_2 r_{\text{out}})}$$

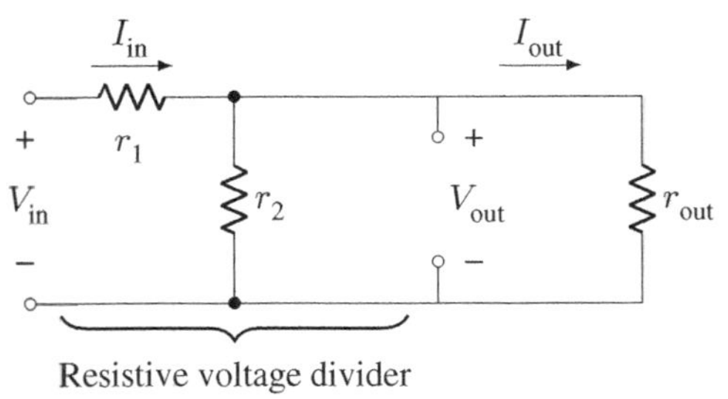

Figure 4.39 Resistive voltage divider circuit for Problem 4.1.

4.2 Show that the compact Fourier series of the symmetric square wave switching function shown in Figure 4.40 is the following:

$$q(t) = \frac{1}{2} + \frac{2}{\pi} \sum_{n=1,3,5,\ldots}^{\infty} \frac{\sin(nt)}{n}$$

It may help to compare $q(t)$ to Figure 4.3, and use (4.3).

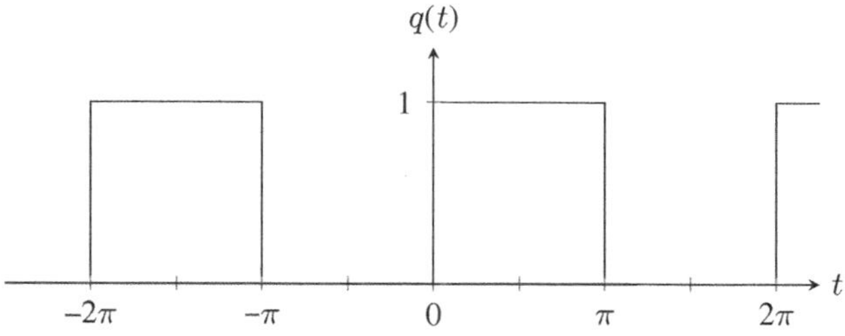

Figure 4.40 Symmetric square wave switching function for Problem 4.2.

4.3 In this problem, we will derive the compact Fourier series for the periodic triangle wave *tri*(t) shown in Figure 4.41. *tri*(t) has period T and is defined over $[0, T)$ as

$$tri(t) \triangleq \begin{cases} 1 - \frac{4t}{T}, & \forall t \in [0,\ T/2) \\ -1 + \frac{4(t - T/2)}{T}, & \forall t \in \left[T/2,\ T\right) \end{cases}$$

(a) If we define a function $f(t) \triangleq \frac{d}{dt} tri(t)$, what is $f(t)$? Describe and/or sketch the function or write it in functional form like the definition of *tri*(t) above.

(b) Compare your answer for $f(t)$ in (a) to $q(t)$ in (4.3) and Figure 4.3. If $f(t) = \alpha q_0(t - x) + \beta$, where $q_0(t) \triangleq q(t)|_{\theta(0)=0}$ (i.e. $f(t)$ is a scaled and shifted version of the generalized switching function with $\theta(0) = 0$), find α, x, and β.

(c) Evaluate (4.3) at $\theta(0) = 0$ to obtain $q_0(t)$, then use the result in (b) to show that the Fourier series representation of $f(t)$ is the following [*Hint*: you will find this exercise similar to SP4.5.1]:

$$f(t) = -\frac{16}{\pi T} \sum_{n=1,3,5,\ldots}^{\infty} \frac{1}{n} \sin\left(\frac{2\pi n t}{T}\right)$$

(d) Use the result in (c) to show that the Fourier series for *tri*(t) is the following:

$$tri(t) = \frac{8}{\pi^2} \sum_{n=1,3,5,\ldots}^{\infty} \frac{1}{n^2} \cos\left(\frac{2\pi n t}{T}\right)$$

You may find the following identity useful: $\sum_{n=1,3,5,\ldots}^{\infty} \frac{1}{n^2} = \frac{\pi^2}{8}$.

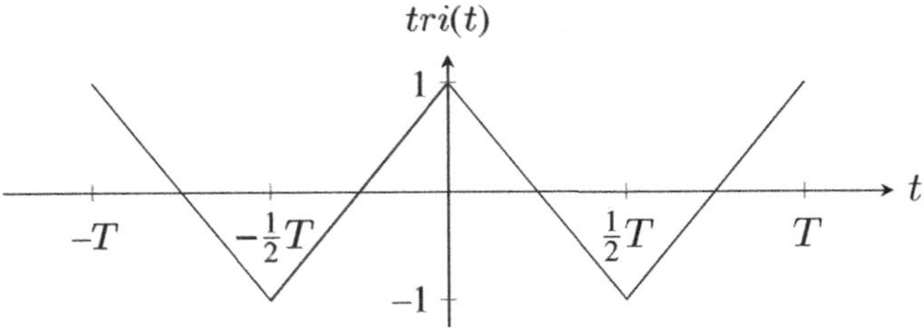

Figure 4.41 Triangle waveform for Problem 4.3.

4.4 Repeat the buck converter design in Example 4C if instead the battery voltage is 24 V and the inductor peak-to-peak current ripple should be no more than 20% of the average current. All other specifications and assumptions are the same. Note that your results can vary from the solution, depending on your design choices.

4.5 For the buck converter you designed in Problem 4.4, sketch the inductor current waveform over three switching periods. Be sure to label key points on the *x*- and *y*-axes to explain your sketch.

4.6 You wish to interface a photovoltaic array, which produces dc current at 90 V, to a dc 270 V, 10.8 kW load. (a) Suggest dc–dc converter for this application. (b) Based on your choice of converter, find *k*, assuming no losses. (c) Calculate the load current. (d) Assuming no losses in the converter, calculate the photovoltaic array output current and power.

4.7 Show that the boost conversion ratio for the non-ideal boost converter shown in Figure 4.42 is given by $\frac{\hat{v}_{out}}{V_{in}} = \frac{(1-k)r}{(1-k)^2 r + r_L}$. This result produces the lossy curve in Figure 4.16. [*Hint*: First, write the power balance equation relating average input, output, and loss power. Then, express each power in terms of the average inductor current, $\hat{i}_L$, and other circuit values.]

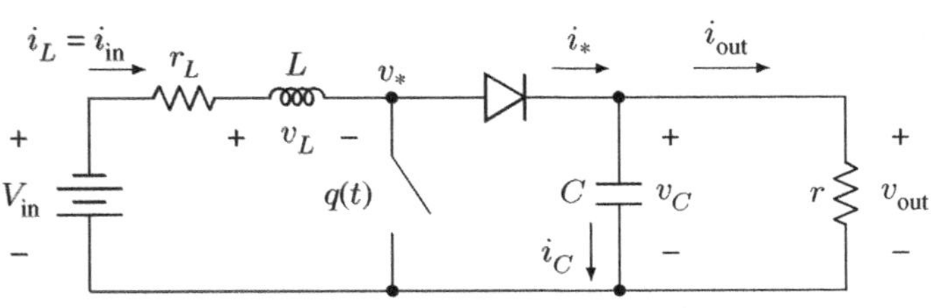

Figure 4.42 Boost converter circuit with resistive losses in the inductor; for Problem 4.7.

4.8 For the diode circuit in Figure 4.43 with input voltage $v_{in}(t)$ shown in Figure 4.44, perform the diode trial method by completing the following steps: (a) Create and fill in a trial method table similar to Table 4.1. (b) Based on your results in (a), sketch $v_{load}(t)$ over a few periods. (c) Calculate $\hat{P}_{load}$.

Figure 4.43 Diode circuit for Problem 4.8.

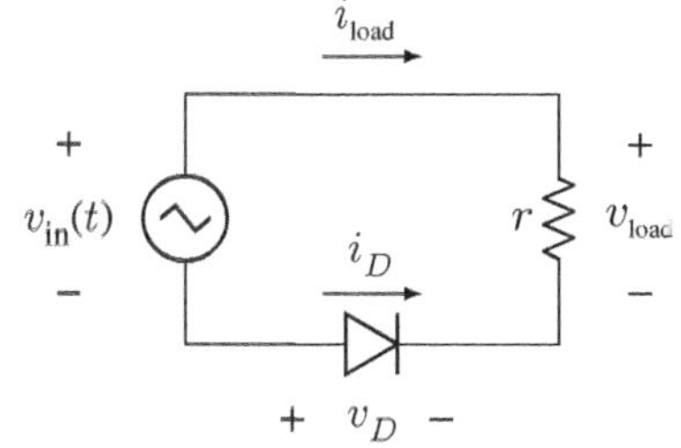

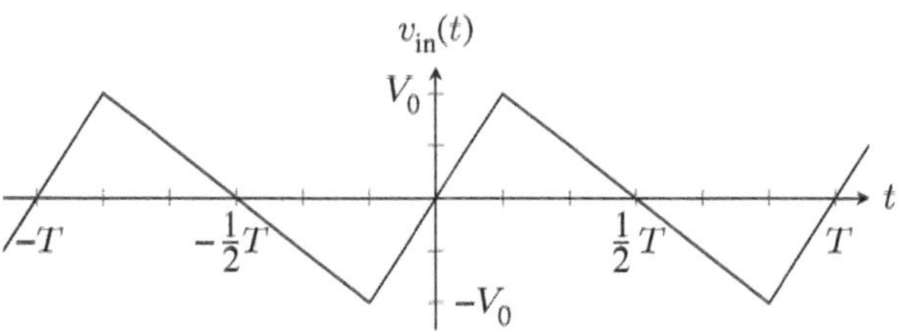

Figure 4.44 Input voltage for diode circuit; for Problem 4.8.

4.9 Conduct the diode trial method and complete a table like Table 4.1 for the full-wave rectifier. [*Hint*: only the states shown in Table 4.2 are viable. Thus, your table will fill in all the un-viable states.]

4.10 A single-phase inverter with $V_{in} = 10\,V$ is supplying a series *rL* load with $r = 1\,\Omega$ and $L = 3\,mH$. The switching frequency, f_{sw}, is 60 Hz and relevant circuit waveforms are shown in Figure 4.37. Find the load current phasor and average power for (a) the fundamental, and (b) the third harmonic. The series you found in SP4.5.1 will be useful.

References

1 Krein, P.T. (1998). *Elements of Power Electronics*. Oxford University Press.

2 Erickson, R.W. and Maksimovic, D. (2001). *Fundamentals of Power Electronics*, 2e. Springer.

3 Mohan, N., Undeland, T.M., and Robbins, W.P. (1989). *Power Electronics: Converters, Applications, and Design*. Wiley.

5

Electric Drives

5.1 Introduction

This chapter provides an introduction to modern electric drives, including the direct current (dc), brushless dc, and induction motor drives. The principle of each torque control is to express the torque in the form of a current. In this way, we can control the commanded torque by controlling the current. The permanent-magnet dc drive is naturally adapted to this type of control. In the case of the permanent-magnet alternating current (ac) machine and the induction machine, both with field orientation, we use synchronous reference frame variables to establish torque-current relationship. This is not an in-depth treatment but rather a brief consideration of each drive sufficient to point out the application of each and to give the student some insight to the drives area. Control theory is not needed to follow this approach. With the exception of the dc drive, the focus is on what each control does, rather than how it does it. In the case of ac drives, reference frame theory is used in the derivation of the steady-state equations; however, it is not necessary to work with these equations. It is understood that this approach is meant only to introduce these drives and the actual study of the control methods is well beyond this introductory treatment. The last subsection in each of the following sections serves as an example as well as being an important part of each section.

5.2 dc Drive

Although the dc machine is not used as extensively as in the past, it still plays a role in some drive applications. A brief look at a method of voltage control is appropriate. Our focus will be on the permanent-magnet dc machine supplied from a two-quadrant dc converter. Dynamic and steady-state performances are illustrated. Since the dc converters used in dc drive systems are often called choppers,

Introduction to the Analysis of Electromechanical Systems, First Edition.
Paul C. Krause, Oleg Wasynczuk, and Timothy O'Connell.

we will use dc converter and chopper interchangeably. In this section, we will analyze the operation and establish the average-value model for a two-quadrant chopper drive. The details of the control may be skipped and study resumed with Section 5.2.1.

A two-quadrant dc converter is depicted in Figure 5.1. The switches $S1$ and $S2$ are transistors. They are assumed to be ideal; that is, if $S1$ or $S2$ is closed, current is allowed to flow in the direction of the arrow; current is not permitted to flow opposite to the arrow. If $S1$ or $S2$ is open, current is not allowed to flow in either direction regardless of the voltage across the switch. If $S1$ or $S2$ is closed and the current is positive, the voltage drop across the switch is assumed to be zero. Similarly, the diodes $D1$ and $D2$ are ideal. Therefore, if the diode current i_{D1} or i_{D2} is greater than zero, the voltage across the diode is zero. The diode current can never be less than zero.

Waveforms of the converter variables during steady-state operation are shown in Figure 5.2. Therein, the switching period T is made large relative to the armature time constant τ_a for the purpose of depicting the transient of the armature current. Normally, the switching period is much smaller than the armature time constant and the switching segments of i_a are nearly sawtooth in shape. This is portrayed later in this section. With a two-quadrant chopper, the armature voltage cannot be negative ($v_a \geq 0$); however, the armature current can be positive or negative. That is, both I_1 and I_2 (Figure 5.2) can be positive, or I_1 can be negative and I_2 positive, or both I_1 and I_2 can be negative. In Figure 5.2, I_1 is negative and I_2 is positive and the average value of i_a is positive. The mode of operation depicted is motor action if ω_r is positive corresponding to counterclockwise (ccw) rotation.

During interval A, $S1$ is closed and $S2$ is open and, at the start of interval A, $i_a = I_1$, which is negative. Since $S2$ is open, a negative i_a (I_1) can only flow through

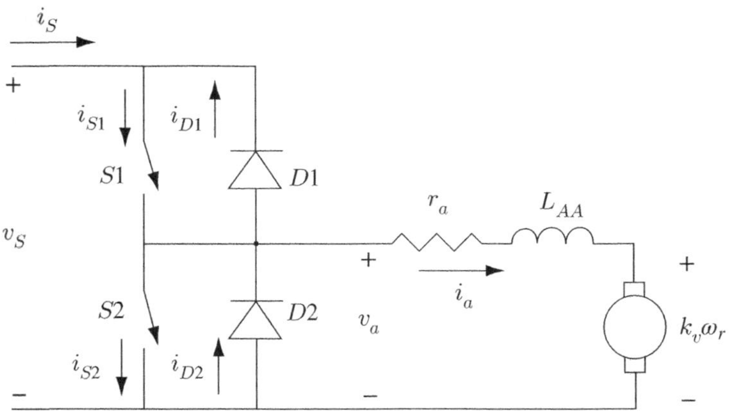

Figure 5.1 Two-quadrant chopper drive.

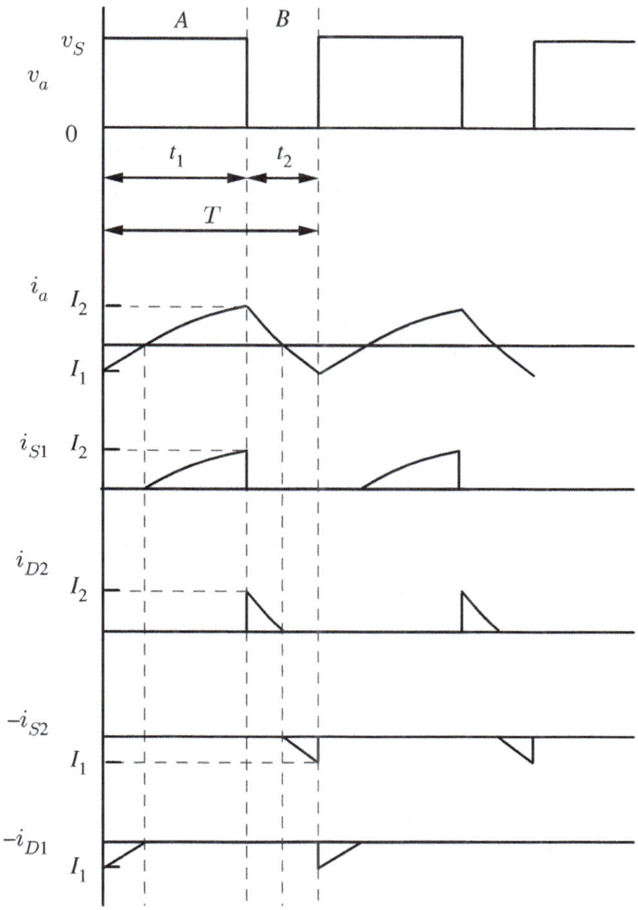

Figure 5.2 Steady-state operation of a two-quadrant dc converter drive.

$D1$. It is important to note that $-i_{D1}$ and $-i_{S2}$ are plotted in Figure 5.2 to allow ready comparison with the waveform of i_a, since they are opposite to positive i_a. Let us go back to the start of interval A. How did i_a become negative? Well, during interval B in the preceding period, $S2$ was closed with $S1$ open. With $S2$ closed, the armature terminals are short-circuited and the counter electromotive force (emf) has driven i_a negative. Therefore, when $S1$ is closed and $S2$ is opened at the start of interval A, the source voltage has to contend with this negative I_1. We see from Figure 5.2 that the average value of i_a is slightly positive; therefore, v_S is larger than the counter emf and at the start of interval A when v_S is applied to the machine the armature current begins to increase toward zero from the negative value of I_1. Once i_a reaches zero, the diode D_1 blocks the current flow. That is, i_{D1} cannot become negative (cannot conduct positive i_a); however, $S1$ has been closed since the start of interval A and since i_{S1} can only be positive, $S1$ is ready to carry the positive i_a. The armature current, which is now i_{S1}, continues to increase until the end of interval A (I_2).

At the beginning of interval B, $S1$ is opened and $S2$ is closed; however, $S2$ cannot conduct a positive armature current. Therefore, the positive current (I_2) is diverted to diode $D2$, which is short-circuiting the armature terminals. Now, the counter emf has the positive current (I_2) with which to contend. It is clear that if the armature terminals were permanently short-circuited, the counter-emf would drive i_a negative. At the start of interval B, the counter-emf begins to do just that; however, when i_a becomes zero, diode $D2$ blocks i_{D2} and the negative armature current is picked up by $S2$, which has been closed since the beginning of interval B, waiting to be called upon to conduct a negative armature current. This continues until the end of interval B, whereupon we are back to where we started.

It is apparent that if the mode of operation is such that both I_1 and I_2 are positive, then the machine is acting as a motor with a substantial load torque if ω_r is positive (ccw). In this mode, either $S1$ or $D2$ will carry current during a switching period T. If both I_1 and I_2 are negative, the machine is operating as a generator, delivering power to the source if ω_r is positive (ccw). In this case, either $S2$ or $D1$ will carry current during a switching period.

5.2.1 Average-value Time-domain Block Diagram

The average-value time-domain block diagram for the two-quadrant chopper drive system is shown in Figure 5.3. From Figure 5.2, the average armature voltage may be determined as

$$\overline{v}_a = \frac{1}{T}\left[\int_0^{t_1} v_S\,d\xi + \int_{t_1}^{T} 0\,d\xi\right] \tag{5.1}$$

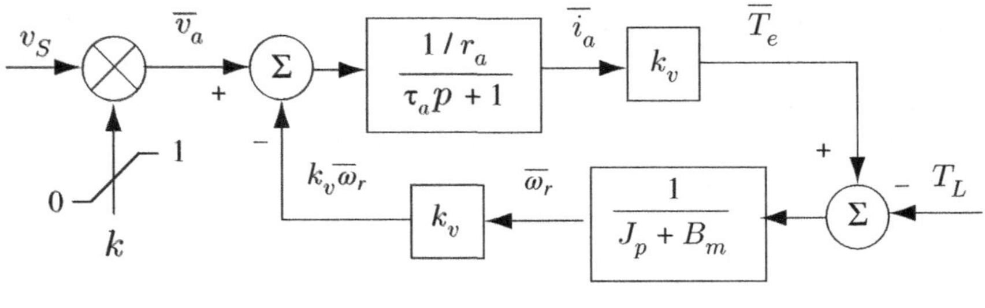

Figure 5.3 Average-value model of two-quadrant dc converter drive.

Since $t_1 = kT$, where k is referred to as the duty cycle, the average armature voltage becomes

$$\bar{v}_a = kv_S \tag{5.2}$$

In Figure 5.3, the bars over the variables denote average values.

The starting characteristics of a permanent-magnet dc machine with a two-quadrant chopper drive are depicted in Figure 5.4. The machine parameters are $r_a = 7\ \Omega$, $L_{AA} = 120$ mH, $k_v = 1.41 \times 10^{-2}$ V s/rad, and $J = 1.06 \times 10^{-6}$ kg m^2; rated voltage is 6 V. Here, the switching frequency f_s is set to 200 Hz and the source voltage to 10 V. Typically, the switching frequency is much higher, generally greater than 20 kHz. The frequency was selected to illustrate the dynamics introduced by the converter. Even at this low switching frequency, the switching period T is much less than the armature time constant τ_a. Thus, the armature current essentially consists of piecewise linear segments about an average response. In Figure 5.4, the duty cycle is stepped from 0 to 0.6, corresponding to a step increase in average applied voltage from 0 to 6 V. The start-up response established using the average-value model is superimposed for purposes of comparison. As shown, the only salient difference between the two responses is

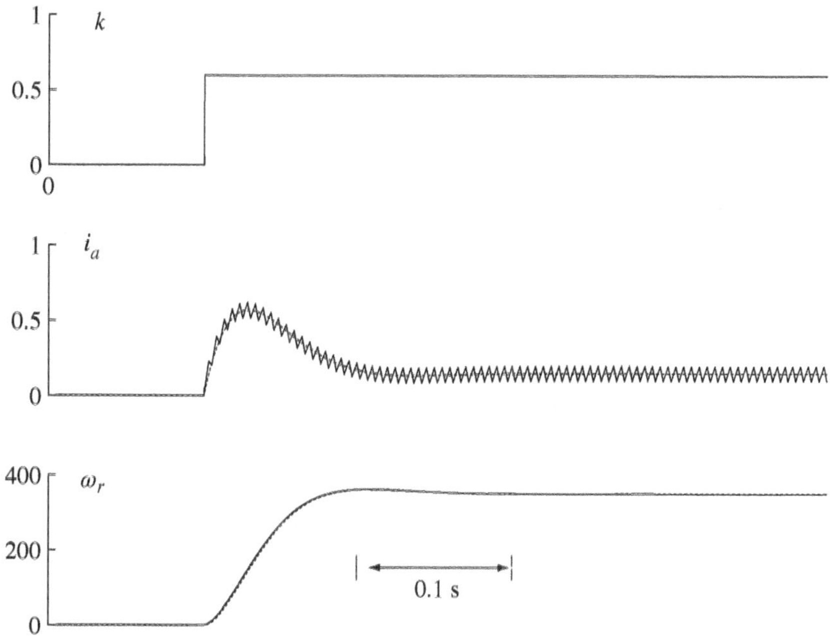

Figure 5.4 Starting characteristics of a permanent-magnet dc machine with a two-quadrant dc/dc converter drive.

the "sawtooth" behavior of the armature current due to converter switching. The rotor speeds are indistinguishable.

5.2.2 Torque Control

The parameters of a permanent-magnet dc machine are $V_a = 6$ V rated, $r_a = 7\ \Omega$, $k_v = 1.41 \times 10^{-2}$ V s/rad, $L_{AA} = 120$ mH, $J = 1.06$ kg m^2, and $B_m = 6.04 \times 10^{-6}$ N m s. We are to limit the torque T_e^* to 0.423×10^{-2} N m or $I_a^* = 0.3$ A where the asterisk denotes commanded values. The control is shown symbolically in Figure 5.5.

Since the current is controlled the electric dynamics are neglected, therefore only the mechanical dynamics are considered. The equations involved in Figure 5.5 are

$$V_a = r_a I_a^* + \omega_r k_v \tag{5.3}$$

$$T_e^* = J\frac{d\omega_r}{dt} + B_m\omega_r + T_L \tag{5.4}$$

where $T_e^* = k_v I_a^* = 0.423 \times 10^{-2}$ N m. The load line is

$$T_L = K\omega_r^2 \tag{5.5}$$

where

$$K = 5.529 \times 10^{-8} \text{ N m s}^2 \tag{5.6}$$

This intersects rated V_a torque versus rotor speed plot at Operating Point 1 where $\omega_r = 276.6$ rad/s, as shown in Figure 5.6.

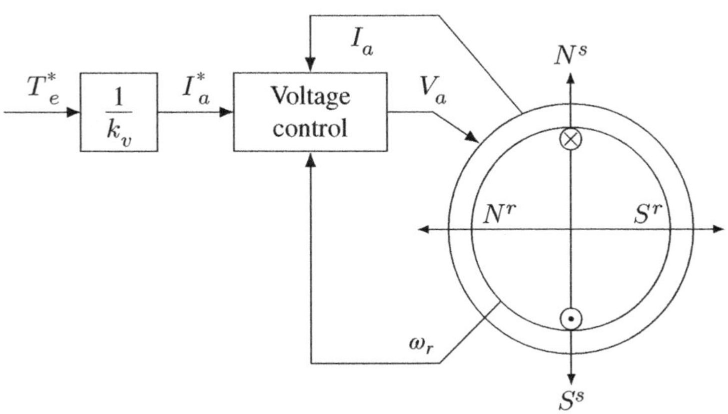

Figure 5.5 Torque control.

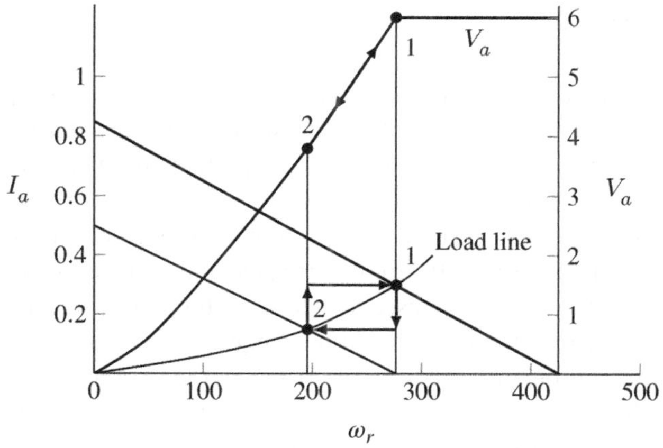

Figure 5.6 Drive operation during T_e^* switching.

The dc machine is operating at Point 1. The commanded torque is suddenly switched to $\frac{1}{2}$ the original value, which intersects the limiting torque (I_a^*) at Operating Point 2 where $\omega_r = 195.6$ rad/s. The electromechanical dynamics are

$$T_e^* = J\frac{d\omega_r}{dt} + B_m\omega_r + T_L \tag{5.7}$$

Assuming the torque control is functioning perfectly, the rotor slows and steady state is reached at Operating Point 2. The voltage at Operating Point 2 is

$$\begin{aligned} V_a &= r_a I_a^* + k_v\omega_r \\ &= (7)(0.15) + (1.41 \times 10^{-2})(195.6) \\ &= 1.05 + 2.76 = 3.8 \text{ V} \end{aligned} \tag{5.8}$$

The commanded torque T_e^* is returned to the original value. The rotor speeds up and reaches steady state at Operating Point 1. The trajectory from Operating Point 1 to Operating Point 2 and then back to 1 is shown in Figure 5.6.

SP5.2.1 The dc machine given in torque control is supplied from a two-quadrant converter. Determine k to reduce the no-load speed to one-half that when full voltage is applied to the armature. [$k = 0.5$]

SP5.2.2 Sketch Figure 5.3 for only motor action; i.e. i_{S2} and i_{D1} equal to zero.

5.3 Brushless dc Drive

The brushless dc drive that we will consider consists of the three-phase six-step inverter and the permanent-magnet ac machine. The six-step inverter and the switching logic are shown in Figure 5.7. The three-phase permanent-magnet ac machine is shown in Figure 5.8. The three sensors shown in Figure 5.8 are generally Hall-effect devices; when the south pole of the permanent-magnet rotor is under a sensor its output is nonzero; with the rotor north pole under the sensor, its output is zero. The states of the sensors are used to determine the switching logic for the inverter, which, in turn, determines the output frequency of the inverter. In the actual machine, the sensors may not be positioned over the rotor as shown in Figure 5.8. Instead, they are placed over a ring that is mounted on the shaft external to the stator and magnetized the same as the rotor.

Since $v_{as} + v_{bs} + v_{cs} = 0$, from Figure 5.7a and Example 1D, we can write

$$v_{as} = \frac{2}{3}v_{ag} - \frac{1}{3}(v_{bg} + v_{cg}) \tag{5.9}$$

$$v_{bs} = \frac{2}{3}v_{bg} - \frac{1}{3}(v_{cg} + v_{ag}) \tag{5.10}$$

$$v_{cs} = \frac{2}{3}v_{cg} - \frac{1}{3}(v_{ag} + v_{bg}) \tag{5.11}$$

In (5.9)–(5.11), v_{ag}, v_{bg}, and v_{cg} are either V_{in} or zero depending upon the state of transistor switches S_1–S_6.

Now, let us write the displacement of the voltage as

$$\phi_v = \theta_{esv} - \theta_r \tag{5.12}$$

where, in the steady state

$$\theta_{esv} = \omega_e t + \theta_{esv}(0) \tag{5.13}$$

$$\theta_r = \omega_r t + \theta_r(0) \tag{5.14}$$

The voltage v_{as} may be changed in magnitude by changing V_{in} or by changing the phase θ_{esv}, since we will select the time-zero position of the q axis to be at $\theta_r = 0$, thus, $\theta_r(0) = 0$. The six-step *as*-phase voltage may be written

$$v_{as} = \frac{2}{\pi}V_{\text{in}}\left[\cos\theta_{esv} + \frac{1}{5}\cos 5\theta_{esv} - \frac{1}{7}\cos 7\theta_{esv} + \cdots\right] \tag{5.15}$$

where ω_e is controlled to be ω_r.

It should be noted that the maximum value of v_{as} given in Figure 5.7c is $\frac{2}{3}V_{\text{in}}$; however, the peak value of the fundamental component of v_{as} given in (5.15) is $\frac{2}{\pi}V_{\text{in}}$. The voltages v_{bs} and v_{cs} may be expressed, for an *abc* sequence, by substituting $\theta_{esv} - \frac{2}{3}\pi$ and $\theta_{esv} + \frac{2}{3}\pi$, respectively, for θ_{esv} in (5.15).

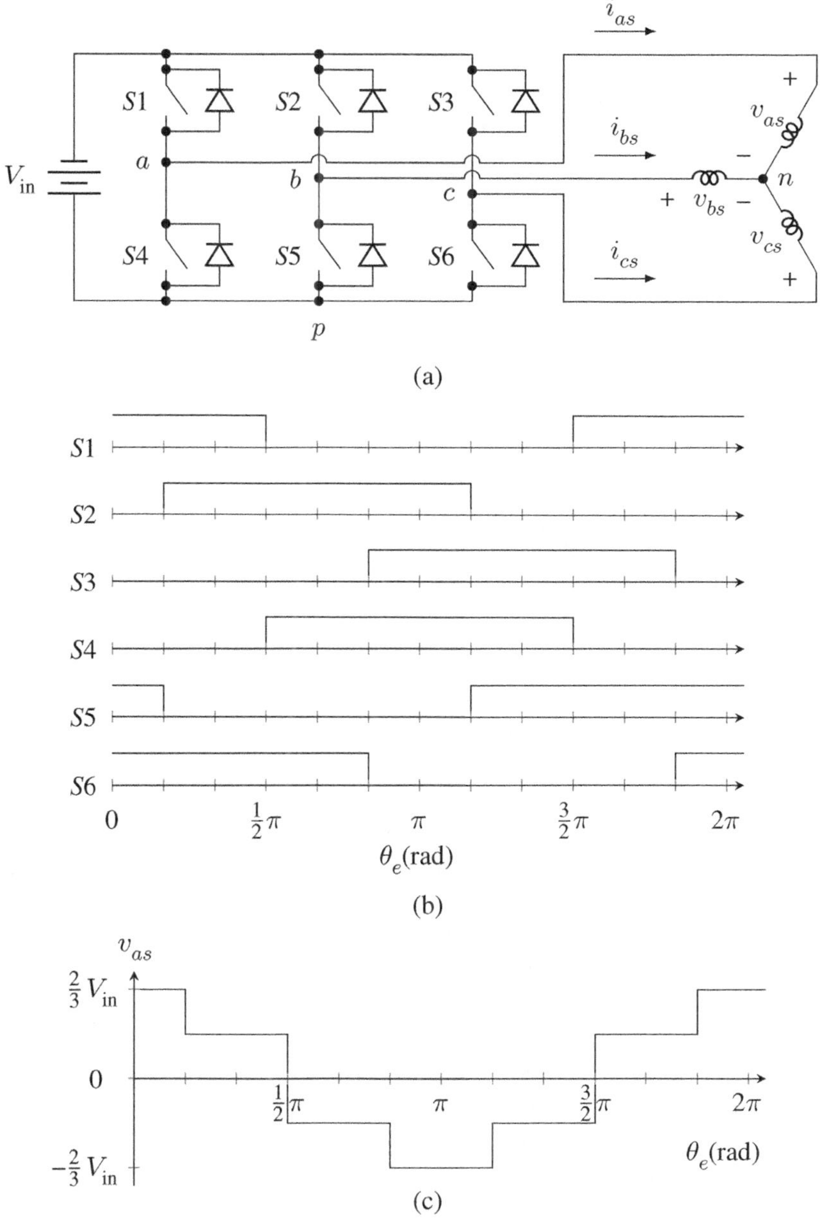

Figure 5.7 Inverter-machine drive. (a) Inverter configuration, (b) transistor switching logic, and (c) plot of v_{as}.

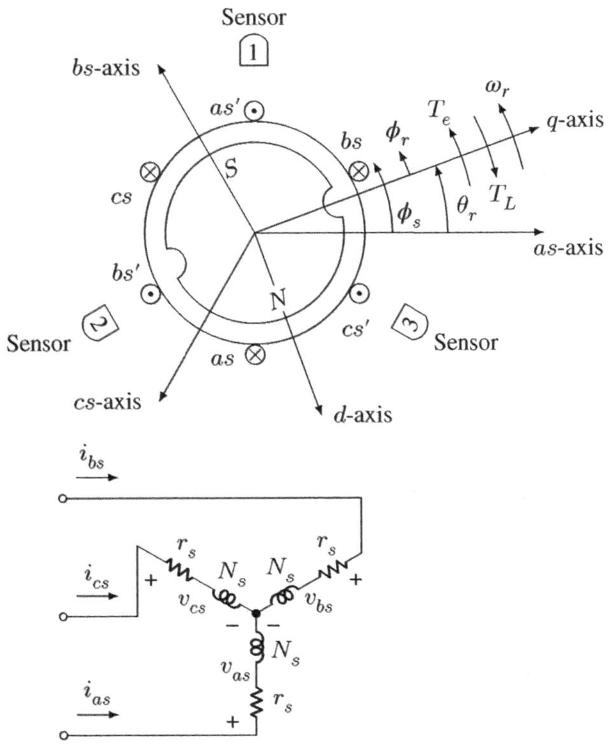

Figure 5.8 Two-pole three-phase permanent-magnet ac machine with sensors.

Before getting into the operation of the brushless dc drive, let us take a closer look at the voltage and current waveforms of the continuous-current inverter. For this purpose, it is sufficient to focus on v_{as} and i_{as} shown in Figure 5.9. We will work with Figures 5.7a and 5.9. In Figure 5.9, v_{as}, i_{as}, i_{aS1}, $-i_{aD1}$, $-i_{aS4}$, and i_{aD4} are plotted for a typical operating condition with a motor load, where i_{aD1} (i_{aD4}) is the current flowing in the anti-parallel diode of $S1$ ($S4$). The diodes are not labeled in Figure 5.7a. Let us start at the center of the peak value of v_{as}. We see from Figure 5.7a that at this instant $S1$ is carrying positive i_{as} and it is being returned to the source through $S5$ (i_{bs}) and $S6$ (i_{cs}) (not shown in Figure 5.9). At the first step $S2$ is turned on, $S5$ is turned off, and v_{as} drops to $\frac{1}{3}V_{\text{in}}$. Now, $S1$ (i_{as}) and $S2$ (i_{bs}) (not shown in Figure 5.9) will share the positive current and $S6$ (i_{cs}) will return the current to the source. At the next step, $S1$ is turned off and $S4$ is turned on. Now $S2$ (i_{bs}) has all of positive current and $S4$ (i_{as}) and $S6$ (i_{cs}) will share the return current. However, since we have an inductive circuit, before $S4$ can share the return current there must be a path for the positive current that was flowing in $S1$ (i_{as}) at

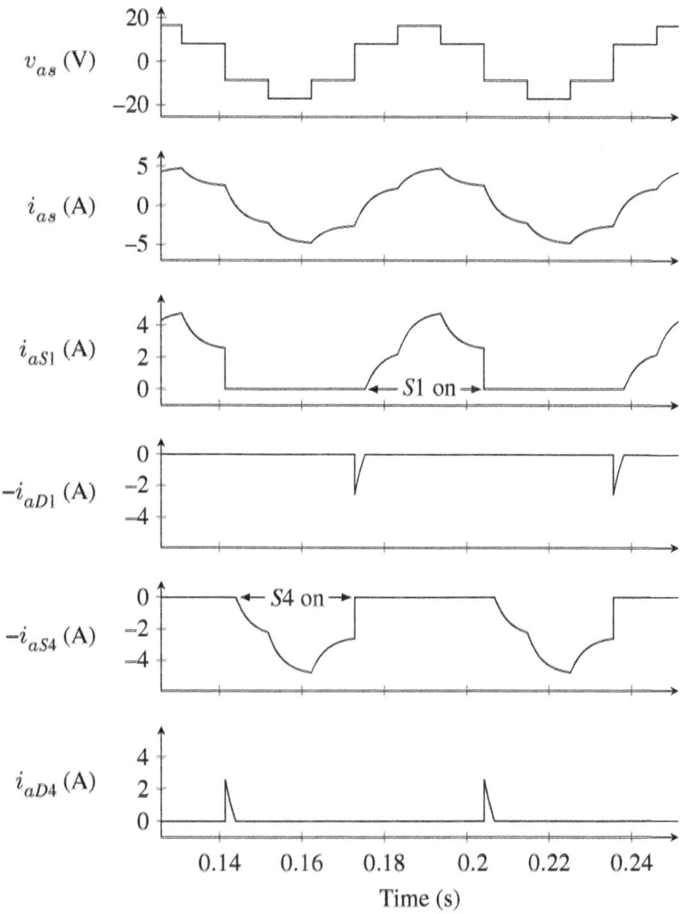

Figure 5.9 Plots of v_{as} and i_{as} and the components of i_{as}.

the time it was turned off. This is through i_{D4} (the anti-parallel diode of $S4$). When $S4$ is turned off the anti-parallel diode of $S1$ springs into action. The pattern is clear.

5.3.1 Operation of Brushless dc Drive with $\phi_v = 0$

The free-acceleration characteristic of this brushless dc motor-inverter drive with $\phi_v = 0$ is shown in Figure 5.10. When $\phi_v = 0$, $\widetilde{V}_{as}$ is located on the q axis along with $\widetilde{E}_a$. The machine parameters are those given previously and repeated here as a four-pole machine with $r_s = 3.4\,\Omega$, $L_{ls} = 1.1\,\text{mH}$, $L_{ms} = 11\,\text{mH}$, and $\lambda'_m = 0.0826\,\text{V s/rad}$. For purposes of comparison, the inverter voltage V_{in} is 25 V, which is equal to the value used in the case of sinusoidal applied stator voltages.

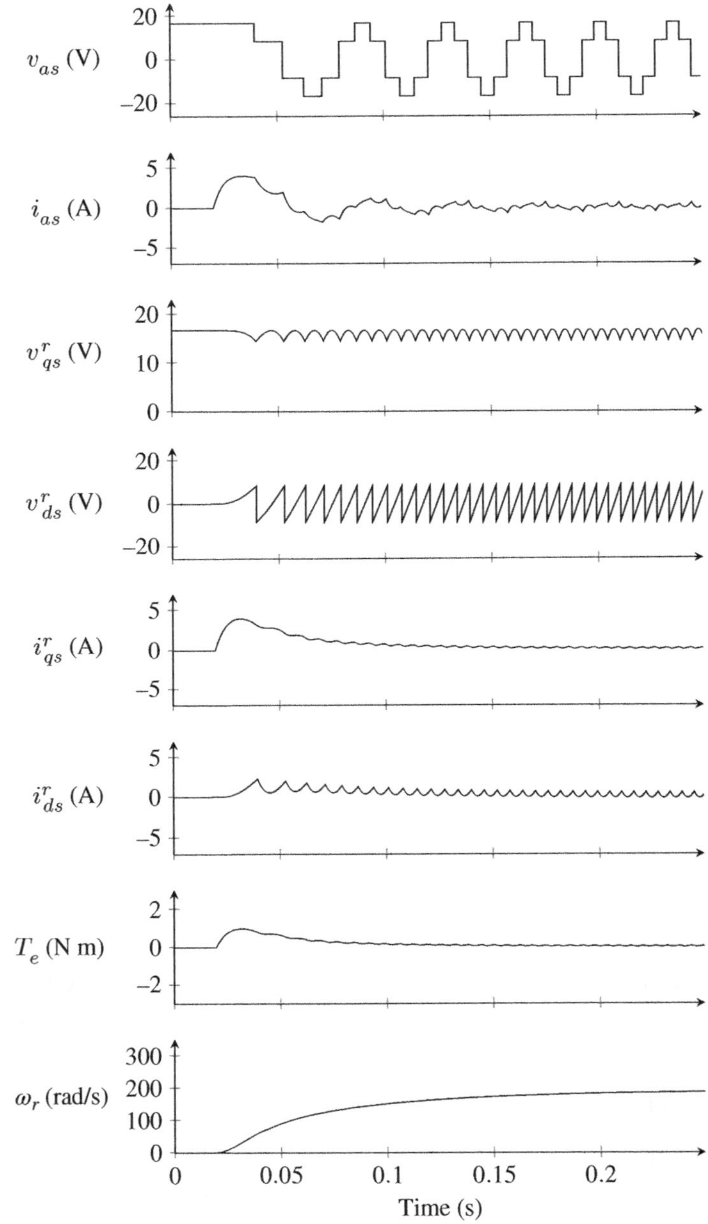

Figure 5.10 Free-acceleration characteristics of a three-phase brushless dc motor supplied by a six-step inverter with $\phi_v = 0$.

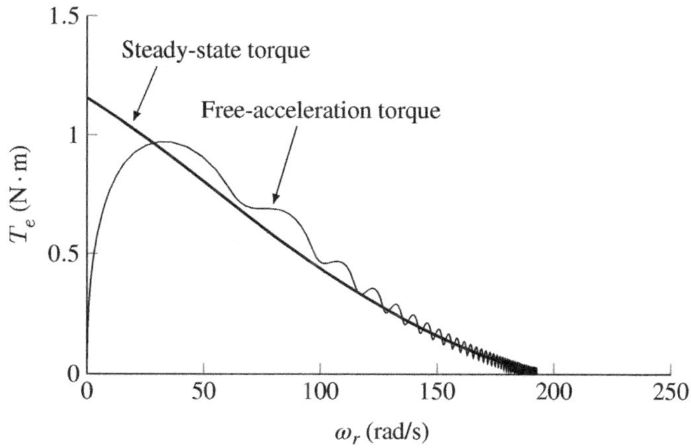

Figure 5.11 Torque-speed characteristics for free acceleration shown in Figure 5.10.

The torque-versus-speed characteristics for this free acceleration are shown in Figure 5.11. In Figure 5.11, we see the reason for the name "brushless dc" since when $\phi_v = 0$ the steady-state torque versus speed characteristics are similar to that of a dc machine. The dynamic performance of this system during load-torque switching is shown in Figure 5.12. The inertia is 2×10^{-4} kg m^2, and the load torque is switched from 0.1 to 0.4 N m and then back to 0.1 N m.

It is instructive to neglect the voltage harmonics and set the voltage equal to the first term of (5.15), that is

$$v_{as} = \frac{2}{\pi} V_{\text{in}} \cos \theta_{esv} \tag{5.16}$$

These plots are shown in Figure 5.13. Neglecting the harmonic content of the voltages is a reasonable approximation for many applications.

There are two other modes of operation of a brushless dc drive that are of importance. In particular, (i) maximum torque per volt, $\phi_{vMT/V}$, and (ii) maximum torque per amp, $\phi_{vMT/A}$. These modes of operation are achieved by shifting the applied voltage relative to $\widetilde{E}_a$ and are covered in [1–3].

5.3.2 Torque Control

The parameters of the four-pole three-phase permanent magnet ac machine are $r_s = 3.4\,\Omega$, $L_{ls} = 1.1$ mH, $L_{ms} = \frac{3}{2}\,11$ mH, $\lambda_m' = 0.0826$ V s/rad, and rated $V_s = 11.25$ V. The torque is controlled with $\phi_v = 0$ at 0.315 N m until rated voltage is reached at $\omega_r = 120.1$ rad/s. The voltage V_s is then maintained at 11.25 V with $\phi_v = 0$. The load line and Operating Point 1 are as shown in Figure 5.14.

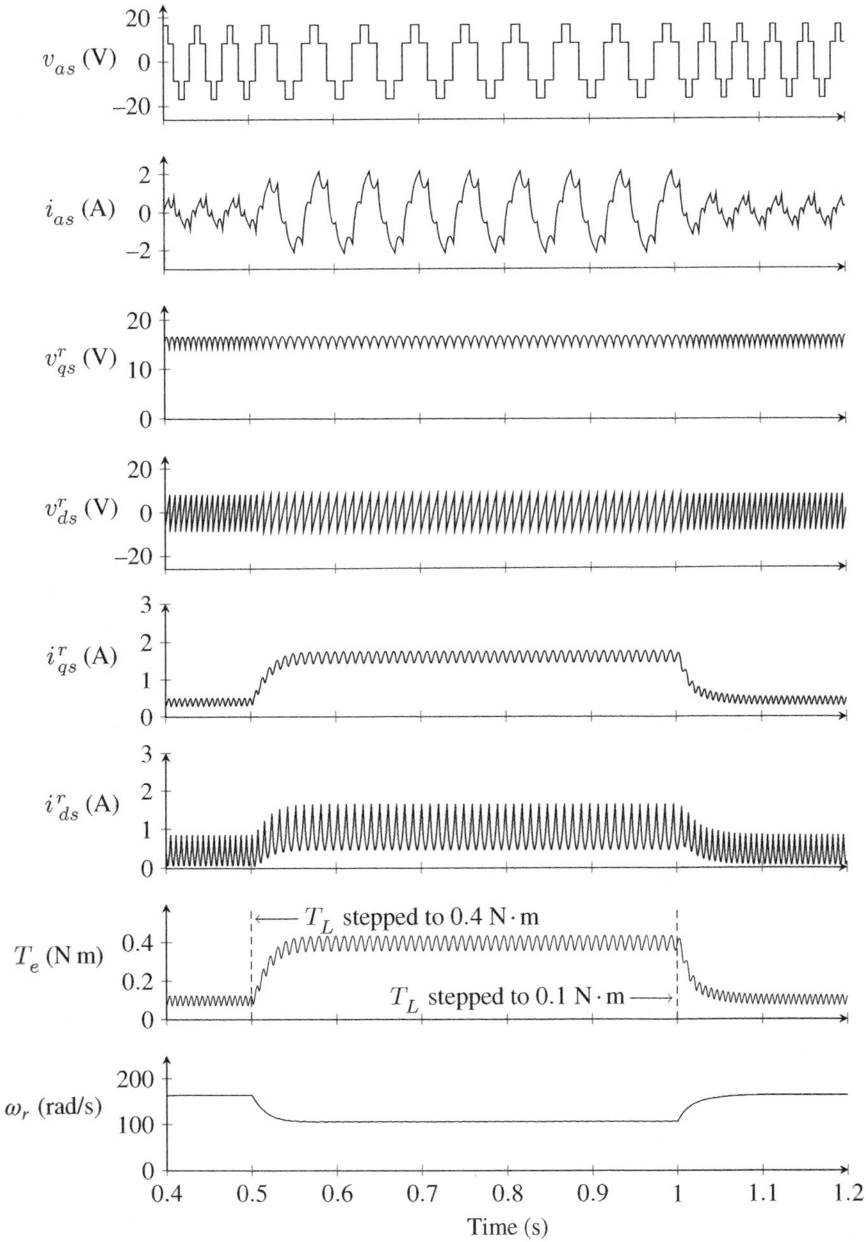

Figure 5.12 Dynamic performance during step changes in load torque of a three-phase, brushless dc motor supplied by an inverter with $\phi_v = 0$.

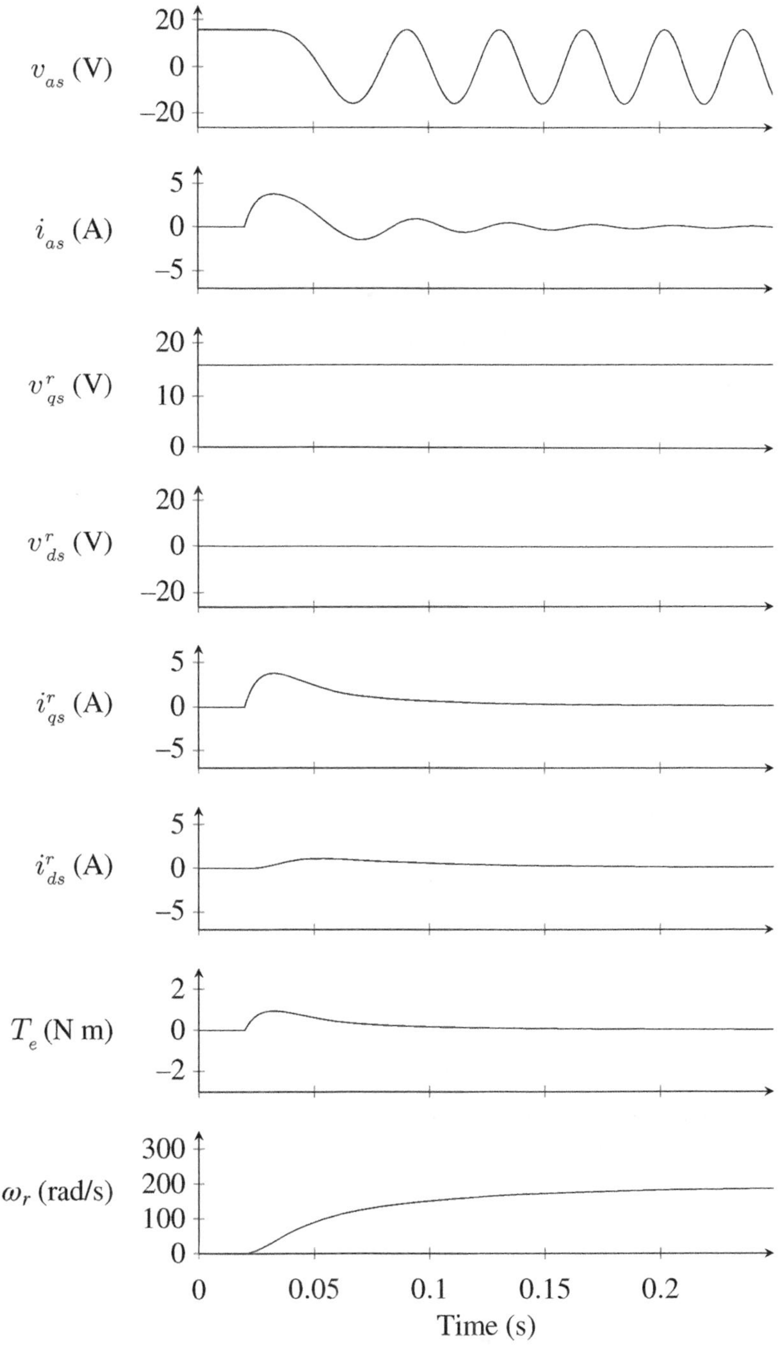

Figure 5.13 Free-acceleration characteristics of a brushless dc drive with harmonics neglected and $\phi_v = 0$ and a total inertia of 5×10^{-4} kg m^2.

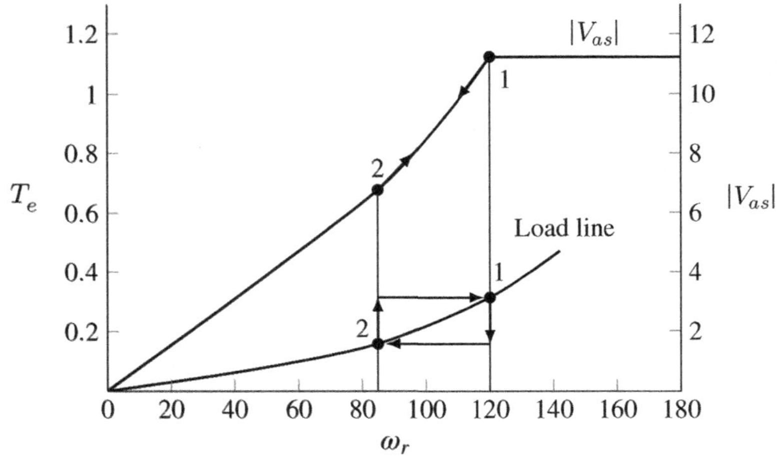

Figure 5.14 Controlling T_e with $\phi_v = 0$.

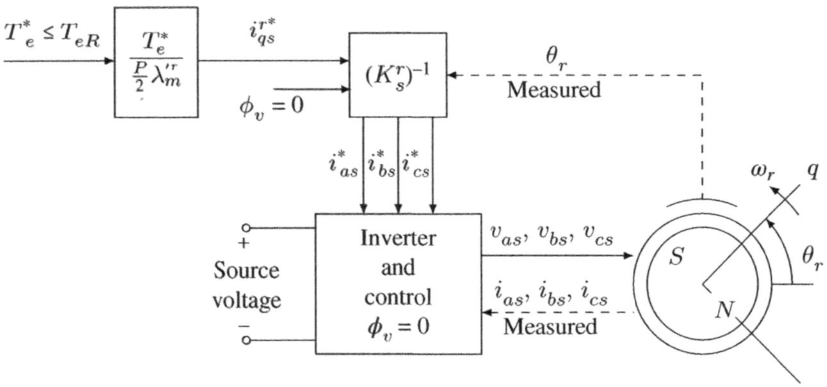

Figure 5.15 Block diagram of torque control of a three-phase permanent-magnet ac machine with $\phi_v = 0$.

Although we will not become involved in the control, the block diagram given in Figure 5.15 is a general top-level view of the control.

The load line is calculated from

$$T_L = K\omega_r^2 \tag{5.17}$$

where $T_L = 0.315$ N m and $\omega_{rm} = 60$ rad/s. Substituting into (5.17) and solving for K yields

$$K = \frac{0.315}{(60)^2} = 8.75 \times 10^{-5} \text{ N m s} \tag{5.18}$$

The commanded torque T_e^* is suddenly reduced to one-half the original. Thus,

$$\begin{aligned} T_e^* &= \left(\frac{1}{2}\right)(0.315) \\ &= 0.1575 \text{ N m} \end{aligned} \tag{5.19}$$

The trajectory from Operating Point 1 to Operating Point 2 is shown in Figure 5.14. The rotor speed at Operating Point 2 can be calculated as

$$T_L = K\omega_{rm}^2 \tag{5.20}$$

where $T_L = 0.1575$ N m and $K = 8.75 \times 10^{-5}$ N m s^2. This yields

$$\omega_r = \frac{P}{2}\left(\frac{0.1575}{8.75 \times 10^{-5}}\right)^{\frac{1}{2}} = 84.9 \text{ rad/s} \tag{5.21}$$

Now from (3.63)

$$T_e = \left(\frac{3}{2}\right)\left(\frac{P}{2}\right)\lambda_m'^r \sqrt{2} I_s \cos\theta_{esi}(0) \tag{5.22}$$

We can write

$$\sqrt{2} I_s \cos\theta_{esi}(0) = \frac{T_e}{\left(\frac{3}{2}\right)\left(\frac{P}{2}\right)\lambda_m'^r} = \frac{0.1575}{\left(\frac{3}{2}\right)\left(\frac{4}{2}\right)0.0826} = 0.636 \text{ A} \tag{5.23}$$

We can calculate the voltage from (3.60); however, we first need $\tilde{I}_{as}$. Since $\phi_v = 0$ $\tilde{V}_{as}$ and $\tilde{E}_a$ are at zero degrees; therefore, the phase angle of $\tilde{I}_{as}$ is

$$\begin{aligned} \theta_{esi}(0) &= -\tan^{-1}\frac{\omega_r L_{ss}}{r_s} \\ &= -\tan^{-1}\frac{(84.9)\left(1.1 + \frac{3}{2}11\right) \times 10^{-3}}{3.4} \\ &= -\tan^{-1} 0.439 = -23.7^\circ \end{aligned} \tag{5.24}$$

From (5.23)

$$I_s = \frac{0.636}{\sqrt{2}\cos 23.7} = 0.4912 \tag{5.25}$$

Therefore,

$$\tilde{I}_{as} = 0.4912\underline{/-23.7^\circ} \tag{5.26}$$

Substituting into (3.60) yields

$$\begin{aligned}\widetilde{V}_{as} &= (r_s + j\omega_r L_{ss})\widetilde{I}_{as} + \frac{1}{\sqrt{2}}\omega_r \lambda_m'^r \underline{/0^\circ} \\ &= \left(3.4 + j(84.9)\left\{\left(1.1 + \frac{3}{2}11\right) \times 10^{-3}\right\}\right)(0.4912\underline{/-23.7^\circ}) \\ &\quad + \frac{(84.9)(0.0826)}{\sqrt{2}} = (3.714\underline{/23.7})(0.4912\underline{/-23.7}) + 4.96 = 6.78 \text{ V}\end{aligned} \tag{5.27}$$

After steady-state operation is reached at Operating Point 2, the commanded torque is increased back to 0.315 N m. The trajectory from Operating Point 2 to Operating Point 1 is shown in Figure 5.14. The electromechanical dynamics are determined by (2.114) which is

$$T_e^* = J\frac{2}{P}\frac{d\omega_r}{dt} + B_m\frac{2}{P}\omega_r + T_L \tag{5.28}$$

where $J = 5 \times 10^{-4}$ kg m^2 and $B_m = 0$.

SP5.3.1 With the initial conditions assumed, the stall torque of an inverter-driven brushless dc machine is slightly larger than the sinusoidal approximation. Why? $\left[\frac{2}{3}V_{\text{in}} \text{ versus } \frac{2}{\pi}V_{\text{in}}\right]$

SP5.3.2 For a given inverter voltage, what factors determine the initial starting torque? [rotor position, position of mmf$_s$]

5.4 Induction Motor Drive

In the previous section, the permanent-magnet ac machine was used in a brushless dc drive where nearly instantaneous current control was achieved by controlling the speed of the stator rotating magnetic field equal to the electrical angular velocity of the rotor. This was accomplished by controlled switching of the drive inverter that supplies the stator applied voltages; however, we did not become involved with the design of the control. Instead, we analyzed the steady-state performance assuming that the sensing and control functions were performing perfectly, whereupon, the control of torque was essentially instantaneous. The electric transients were essentially non-existent; that is, a change from one steady-state operating point to another was accomplished with a minimum of electric transients. Therefore, considering only steady-state performance was not as unrepresentative as one might have first thought.

The goal of field orientation of an induction machine is similar; fast torque response, which is accomplished with a controlled drive inverter [4, 5]. Unfortunately, the necessary control is more involved in the case of field orientation of the induction machine since not only is a torque control involved but it is also necessary to control the stator variables to orient the maximum value of the rotor field, mmf_r^e, in the q axis, orthogonal with the d-axis component of the stator field, mmf_s^e.To do all this we have two things we can control, the stator applied voltage and the slip frequency. Although the principle of field orientation is quite straightforward, the implementation of the control is not. In particular, the voltage equations in the synchronous rotating reference frame are used to determine the values of the substitute currents necessary to ensure field orientation and these values are then transformed from the synchronous reference frame to the stationary reference frame to determine the waveforms of the stator currents and applied voltages necessary to produce the synchronous reference frame variables that will provide field orientation. This is achieved by controlling the output voltage of the drive inverter to shape the waveforms of the actual stator currents to that determined necessary to provide field orientation. In this section, we will not become involved with the details of implementing this control; instead, we will follow the same approach as in the case of the brushless dc drive in that we will assume that the control is functioning perfectly and focus on the steady-state performance of an induction machine with field orientation. In other words, in this section, our focus will be on what the control does and not how it does it. Also, we will consider the two-phase device since this reduces our work and the extension to three-phase is straightforward.

If the reader wishes to bypass the reference-frame derivation of the phasor voltage equation, the material up to (5.47) and (5.48) may be deemphasized. The voltage and flux linkage equations of a two-phase single-fed induction machine in the synchronously rotating reference frame are

$$v_{qs}^e = r_s i_{qs}^e + \omega_e \lambda_{ds}^e + p\lambda_{qs}^e \tag{5.29}$$

$$v_{ds}^e = r_s i_{ds}^e - \omega_e \lambda_{qs}^e + p\lambda_{ds}^e \tag{5.30}$$

$$0 = r_r' i_{qr}'^e + (\omega_e - \omega_r)\lambda_{dr}'^e + p\lambda_{qr}'^e \tag{5.31}$$

$$0 = r_r' i_{dr}'^e - (\omega_e - \omega_r)\lambda_{qr}'^e + p\lambda_{dr}'^e \tag{5.32}$$

where

$$\lambda_{qs}^e = L_{ss} i_{qs}^e + L_{ms} i_{qr}'^e \tag{5.33}$$

$$\lambda_{ds}^e = L_{ss} i_{ds}^e + L_{ms} i_{dr}'^e \tag{5.34}$$

$$\lambda'^{e}_{qr} = L'_{rr} i'^{e}_{qr} + L_{ms} i^{e}_{qs} \tag{5.35}$$

$$\lambda'^{e}_{dr} = L'_{rr} i'^{e}_{dr} + L_{ms} i^{e}_{ds} \tag{5.36}$$

The torque may be expressed

$$T_e = \frac{P}{2}\left(\lambda'^{e}_{qr} i'^{e}_{dr} - \lambda'^{e}_{dr} i'^{e}_{qr}\right) \tag{5.37}$$

The aim is to select the applied stator voltages so that $\lambda'^{e}_{qr} = 0$; whereupon, (5.37) becomes

$$T_e = -\frac{P}{2}\lambda'^{e}_{dr} i'^{e}_{qr} \tag{5.38}$$

Please be careful here. Making λ'^{e}_{qr} zero does not mean that i'^{e}_{qr} will also be zero; instead, from (5.35) we see that with λ'^{e}_{qr} zero then

$$i'^{e}_{qr} = -\frac{L_{ms}}{L'_{rr}} i^{e}_{qs} \tag{5.39}$$

Thus, if i'^{e}_{qr} is (5.39) then $\lambda^{e}_{qr} = 0$ and T_e is (5.38).

From (5.38) we see that if λ'^{e}_{dr} is constant then torque is proportional to i^{e}_{qs} from (5.39). We will work toward that goal but first let's control i'^{e}_{dr} to zero. Why? Well, if i'^{e}_{dr} is zero, the rotor poles will be positioned completely in the q axis. Now again, be careful; controlling i'^{e}_{dr} to zero does not mean that λ'^{e}_{dr} is zero; from (5.36) with $i'^{e}_{dr} = 0$ then

$$\lambda'^{e}_{dr} = L_{ms} i^{e}_{ds} \tag{5.40}$$

which will be constant if i^{e}_{ds} is held constant. If (5.39) and (5.40) are substituted into (5.38), the torque may be expressed

$$T_e = \frac{P}{2}\frac{L^2_{ms}}{L'_{rr}} i^{e}_{qs} i^{e}_{ds} \tag{5.41}$$

Note that we have T_e in terms of stator-related currents and if i^{e}_{ds} is held constant T_e is directly proportional to i^{e}_{qs}. This equation is used to control i^{e}_{qs} when T_e is commanded and with i^{e}_{ds} held constant, generally at its rated value.

We have eliminated (5.32) and since λ'^{e}_{qr} is zero, then so must be $p\lambda'^{e}_{qr}$, and if (5.39) is substituted for i'^{e}_{qr} and (5.40) for λ'^{e}_{dr} into (5.31), the angular velocity of the slip may be calculated as

$$(\omega_e - \omega_r)_{\text{calc}} = \frac{r'_r}{L'_{rr}}\frac{i^{e}_{qs}}{i^{e}_{ds}} \tag{5.42}$$

which we will also denote as $(\omega_s)_{\text{calc}}$ for compactness. This equation is used to control slip frequency. Note that (5.42), like (5.41) is in terms of stator-related currents. We have set forth the basic relationships for field orientation control which in effect positions the rotor poles along the q axis (5.39), orthogonal with the d axis (I^e_{ds}) or at $\theta_{esi}(0)$ with I^e_{qs} and I^e_{ds} or $\tilde{I}_{as}$.

Our purpose now is to consider the steady-state performance of an induction motor assuming that the field orientation is functioning properly. Although this idealized approach is an over-simplification of the control challenges involved, it helps to give insight to the basic features of field orientation; however, since the stator currents are commanded (controlled), the electric transients are minimized. Therefore, we'll find that steady-state, "ideally" controlled operation and the actual field orientation drive with harmonics neglected are very similar.

Let's take a minute to express the steady-state torque assuming field orientation is functioning properly. We can express I^e_{qs} and I^e_{ds}, respectively as

$$I^e_{qs} = \sqrt{2}I_s \cos\theta_{esi}(0) \tag{5.43}$$

$$I^e_{ds} = -\sqrt{2}I_s \sin\theta_{esi}(0) \tag{5.44}$$

where in (5.43) and (5.44), $\omega = \omega_e$ and $\theta(0) = 0$. Substituting (5.43) and (5.44) into (5.41) yields the expression for torque with field orientation as

$$\begin{aligned} T_e &= -\frac{P}{2}\frac{L^2_{ms}}{L'_{rr}}\left[\sqrt{2}I_s \cos\theta_{esi}(0)][\sqrt{2}I_s \sin\theta_{esi}(0)\right] \\ &= -\frac{P}{2}\frac{L^2_{ms}}{L'_{rr}} I^2_s \sin 2\theta_{esi}(0) \end{aligned} \tag{5.45}$$

With field orientation and harmonics neglected, the flux linkages in the synchronous reference frame are zero or constants; whereupon, ideally the time rate of change is zero. Therefore, with ideal functioning field orientation $p\tilde{i}_{as}$ is zero and the instantaneous phasor equations become the steady-state phasor equations. The only dynamic feature is the relationship between rotor speed and torque given by (2.114).

We'll work with the steady-state equations and modify V^e_{qs} and V^e_{ds} to account for the results of field orientation action on the I^e_{qr} and I^e_{ds} variables given earlier; thereby, making it necessary to work only with the stator phasor voltage equations. Substituting I'^e_{qr}, in terms of the I^e_{qs} from (5.39), into (5.33) yields

$$\lambda^e_{qs} = \left(\frac{L_{ss}}{L_{ms}} - \frac{L_{ms}}{L'_{rr}}\right) L_{ms} I^e_{qs} \tag{5.46}$$

We can now express steady-state V^e_{qs} and V^e_{ds}. From (5.29) and (5.30) with $p\lambda = 0$, λ^e_{ds} from (5.34), with I'^e_{dr} zero, and (5.46) for λ^e_{qs}, we have

$$V^e_{qs} = r_s I^e_{qs} + \omega_e L_{ss} I^e_{ds} \tag{5.47}$$

$$V^e_{ds} = r_s I^e_{ds} - \omega_e \left(\frac{L_{ss}}{L_{ms}} - \frac{L_{ms}}{L'_{rr}}\right) L_{ms} I^e_{qs} \tag{5.48}$$

Let's take a minute to express $\widetilde{V}_{as}$ for steady-state operation. Substituting (5.47) and (5.48) into (3.108) yields $\sqrt{2}\,\widetilde{V}_{as}$ with field orientation, that is

$$\begin{aligned}\sqrt{2}\widetilde{V}_{as} &= V_{qs}^{e} - jV_{ds}^{e} \\ &= r_s I_{qs}^{e} + \omega_e L_{ss} I_{ds}^{e} - j(r_s I_{ds}^{e} - \omega_e K_L L_{ms} I_{qs}^{e}) \\ &= r_s(I_{qs}^{e} - jI_{ds}^{e}) + \omega_e L_{ss} I_{ds}^{e} + j\omega_e K_L L_{ms} I_{qs}^{e}\end{aligned} \tag{5.49}$$

where

$$K_L = \frac{L_{ss}}{L_{ms}} - \frac{L_{ms}}{L'_{rr}} \tag{5.50}$$

If now we add and subtract $\omega_e K_L L_{ms} I_{ds}^{e}$ on the right-hand side of (5.49), we can express $\widetilde{V}_{as}$ as

$$\widetilde{V}_{as} = (r_s + j\omega_e K_L L_{ms})\widetilde{I}_{as} + \frac{1}{\sqrt{2}}\omega_e(L_{ss} - K_L L_{ms})I_{ds}^{e} \tag{5.51}$$

Substituting (5.44) for I_{ds}^{e} into (5.51) yields

$$\widetilde{V}_{as} = (r_s + j\omega_e K_L L_{ms})\widetilde{I}_{as} - \omega_e(L_{ss} - K_L L_{ms})I_s \sin\theta_{esi}(0) \tag{5.52}$$

Solving (5.41) for I_{qs}^{e} yields

$$I_{qs}^{e} = \frac{2}{P}\frac{L'_{rr}}{L_{ms}^{2}}\frac{T_e}{I_{ds}^{e}} \tag{5.53}$$

We are also going to use (5.42), but let's wait just a minute before we get to that.

Although we are not going to become involved in the details of the field orientation, the block diagram shown in Figure 5.16 depicts the basic principles of the control. Therein, the input variables containing an asterisk are commanded variables. That is, if the machine is to operate with rated load torque and rated frequency, then T_e^{*} would be rated torque and I_{ds}^{e*} would be constant and generally selected to be the value of I_{ds}^{e} for rated conditions and maintained at that value regardless of T_e^{*}. Now, (5.42), (5.53), and (5.54) which we'll get to in a moment, impose field orientation control on the currents, i_{as}^{*} and i_{bs}^{*}. The inverter forms the voltages v_{as} and v_{bs} to ensure i_{as} and i_{bs} are i_{as}^{*} and i_{bs}^{*}, respectively.

We will continue to use reference frame variable notation for compactness; hopefully, this will not be confusing. Thus, we have I_{qs}^{e*} and I_{ds}^{e*} and, from (5.42), we can calculate $\omega_e - \omega_r$, which we are calling $(\omega_s)_{\text{calc}}$ in Figure 5.16 (the slip angular velocity) and/or $(\omega_e - \omega_r)_{\text{calc}}$ since it is calculated from commanded inputs I_{qs}^{e*} and I_{ds}^{e*} and machine parameters. Now $(\omega_s)_{\text{calc}}$ is the slip necessary for field orientation with the commanded stator-related currents. Now, $(\omega_r)_{\text{meas}}$ is the measured electrical angular velocity of the rotor, and it is added to $(\omega_s)_{\text{calc}}$. Thus,

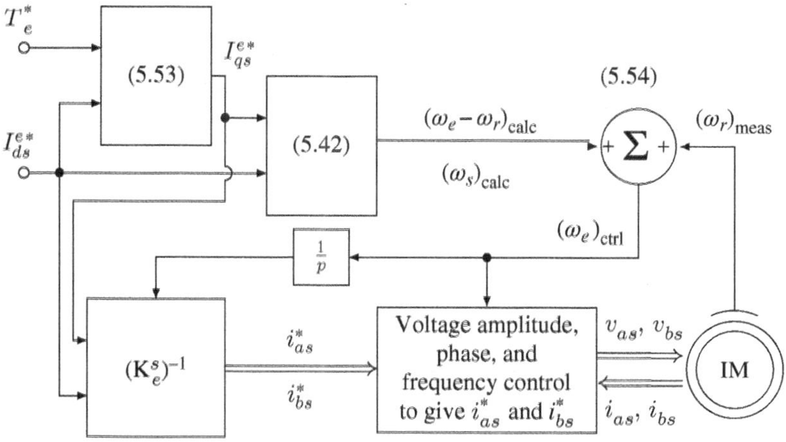

Figure 5.16 Block diagram depicting field-oriented control principles. Note: $I_{qs}^e = \sqrt{2} I_s \cos\theta_{esi}(0)$ and $I_{ds}^e = -\sqrt{2} I_s \sin\theta_{esi}(0)$.

from Figure 5.16

$$(\omega_e)_{ctrl} = (\omega_s)_{calc} + (\omega_r)_{meas}$$
$$= (\omega_e - \omega_r)_{calc} + (\omega_r)_{meas} \tag{5.54}$$

where $(\omega_e)_{ctrl}$ is the electrical angular velocity of the controlled stator voltages applied to the machine (v_{as} and v_{bs} of Figure 5.16) for the rotor magnetic field to be oriented in the q axis. There is no direct control of speed; however, if we are operating in the steady state and the torque load increases, $(\omega_r)_{meas}$ would decrease which would decrease ω_e until $(\omega_e - \omega_r)_{calc}$ was established. Again, $(\omega_e - \omega_r)_{calc}$ is the slip angular velocity, calculated from the commanded T_e^* and I_{ds}^{e*} by (5.42). Assuming the parameters remain constant with temperature change then $(\omega_e - \omega_r)_{calc}$ will change only if T_e^* and/or I_{ds}^{e*} are changed.

5.4.1 Torque Control

The parameters of the two-phase two-pole 5-hp 110-V 60-Hz induction machine are

$r_s = 0.295\ \Omega$	$L_{ms} = 35.15$ mH	$r_r' = 0.201\ \Omega$
$L_{ls} = 0.944$ mH		$L_{lr}' = 0.944$ mH

The machine is equipped with field orientation. The rated torque is 10 N m and the rated power factor angle is 30°. Assume that $\theta_{esi}(0) = -30°$ and $T_e = 10$ N m. This is Operating Point 1 of Figure 5.17 which shows the torque-versus-speed

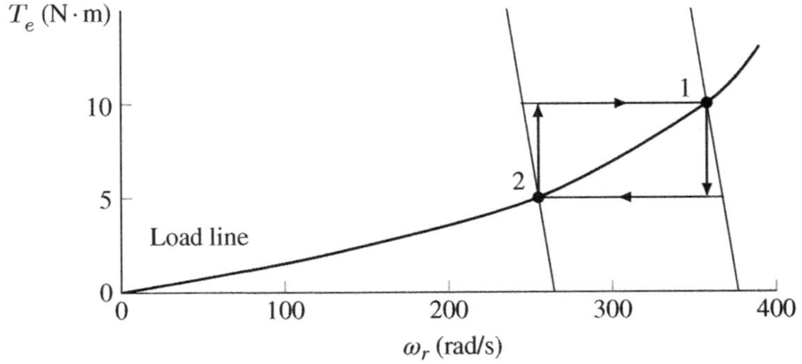

Figure 5.17 Operation of induction motor drive with field orientation for step change in T_e^*.

trajectory for switching the command from 10 to 5 N m and then back to 10 N m. From (5.33)

$$T_e = -\frac{P}{2}\frac{L_{ms}^2}{L'_{rr}} I_s^2 \sin 2\theta_{esi}(0)$$

$$10 = -\left(\frac{2}{2}\right)\frac{(35.15)^2 \times 10^{-3}}{(35.15 + 0.944)} I_s^2 \sin 2(30°) \tag{5.55}$$

Solving for I_s yields

$$I_s = \left[\frac{(10)(35.15 + 0.944)}{(35.15)^2 \times 10^{-3} \sin 60°}\right]^{\frac{1}{2}}$$
$$= 18.37 \text{ A} \tag{5.56}$$

Therefore

$$\tilde{I}_{as} = 18.37\angle -30° \tag{5.57}$$

Now,

$$L_{ss} = L_{ls} + L_{ms}$$
$$= (35.15 + 0.944) \times 10^{-3}$$
$$= 36.094 \text{ mH} \tag{5.58}$$

From (5.50)

$$K_L = \frac{L_{ss}}{L_{ms}} - \frac{L_{ms}}{L'_{rr}} = \frac{36.094}{35.15} - \frac{35.15}{36.094} = 0.0531 \tag{5.59}$$

From (5.51)

$$\tilde{V}_{as} = (r_s + j\omega_e K_L L_{ms})\tilde{I}_{as} - \omega_e(L_{ss} - K_L L_{ms}) I_s \sin\theta_{esi}(0)$$
$$= [0.295 + j377(0.0531)(35.15 \times 10^{-3})]18.37\angle -30°$$
$$- 377(36.094 - 0.0531 \times 35.15) \times 10^{-3}(18.37)(-0.5)$$
$$= 129.7 + j8.5 = 130\angle 3.74° \text{ V} \tag{5.60}$$

This voltage exceeds rated (110 V); however, the current is less than rated. We will continue. The rotor speed is calculated from (5.42)

$$\omega_e - \omega_r = \frac{r'_r}{L'_{rr}} \frac{I^e_{qs}}{I^e_{ds}} \tag{5.61}$$

where

$$\begin{aligned} I^e_{qs} &= \sqrt{2} I_s \cos \theta_{esi}(0) \\ &= \sqrt{2}\, 18.37 \cos(-30°) = 22.5\text{ A} \end{aligned} \tag{5.62}$$

$$\begin{aligned} I^e_{ds} &= -\sqrt{2}\, 18.37 \sin \theta_{esi}(0) \\ &= -\sqrt{2}\, 18.37 \sin(-30°) = 12.99\text{ A} \end{aligned} \tag{5.63}$$

From (5.61)

$$\begin{aligned} \omega_e - \omega_r &= \frac{r'_r}{L'_{rr}} \frac{I^e_{qs}}{I^e_{ds}} \\ &= \frac{0.201}{36.09 \times 10^{-3}} \frac{22.5}{12.99} = 9.6\text{ r/s} \end{aligned} \tag{5.64}$$

$$\begin{aligned} -\omega_r &= -\omega_e + 9.6 \\ \omega_r &= 377 - 9.6 = 367.4\ \text{ r/s} \end{aligned} \tag{5.65}$$

The drive is originally operating at Operating Point 1 of Figure 5.17. The load line is

$$T_L = K\omega_r^2 \tag{5.66}$$

where for $T_e = 10$ N m at Operating Point 1 with $\omega_r = 367.4$ rad/s.

$$K = \frac{10}{(367.4)^2} = 7.41 \times 10^{-5}\text{ N m s}^2 \tag{5.67}$$

The phasor diagram for Operating Point 1 is given in Figure 5.18.

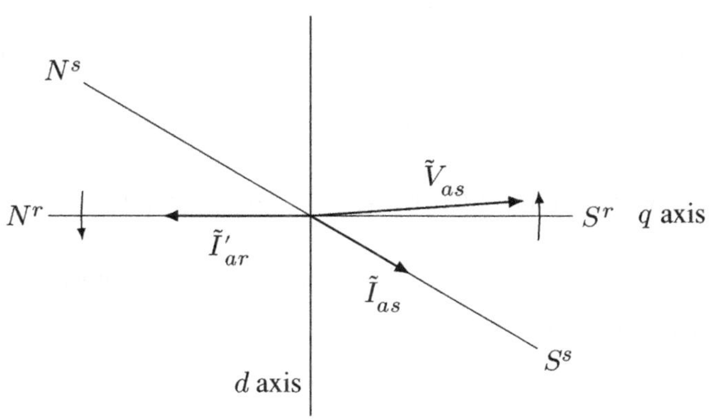

Figure 5.18 Phasor diagram for Operating Point 1.

We see that $\widetilde{I}'_{ar}$ is oriented orthogonal to I^e_{ds} and the torque is 10 N m, and this has been done by controlling $\widetilde{V}_{as}$ and $\omega_e - \omega_r$. The phasor diagram shown in Figure 5.18 should be compared to Figure 3F.2, which is for uncontrolled rotor field orientation.

The commanded torque is changed from 10 to 5 N m. The motor slows and after the mechanical dynamics subside the new Operating Point is at 2 on Figure 5.17. At this point, the new speed from (5.66) is

$$T_L = K\omega_r^2$$
$$5 = 7.41 \times 10^{-5}\omega_r^2$$
$$\omega_r = \left(\frac{5}{7.41 \times 10^{-5}}\right)^{\frac{1}{2}} = 259.8\ \text{rad/s} \tag{5.68}$$

We could liken this scenario to reducing the accelerator of an electric car by one half. Now let us return to Operating Point 1, the accelerator is depressed and we return to Operating Point 1 as shown in Figure 5.17.

Let us calculate the applied voltage of Operating Point 2.

$$\widetilde{V}_{as} = (r_s + j\omega_e K_L L_{ms})\widetilde{I}_{as} - \omega_e(L_{ss} - K_L L_{ms})I_s \sin\theta_{esi}(0) \tag{5.69}$$

Now for $T_e^* = 5$ N m with the same value of I^e_{ds} given by (5.63), $\omega_e - \omega_r = 4.8$ rad/s (see problem 5.7), therefore

$$\omega_e = 259.8 + 4.8 = 264.6\ \text{rad/s} \tag{5.70}$$

Also, $\widetilde{I}_{as} = 12.2\angle{-49.1°}$ and for a two-phase motor $r_s = 0.295\ \Omega$, $L_{ss} = 36.094$ mH, $L_{ms} = 35.15$ mH, and $K_L = 0.053$. Substituting into (5.52)

$$\begin{aligned}\widetilde{V}_{as} &= [0.295 + j264.6(0.053)(35.151 \times 10^{-3})]\ 12.2\angle{-49.1°}\\ &\quad - 264.6(36.094 - 0.053 \times 35.15) \times 10^{-3}(12.2)\sin(-49.1°)\\ &= 0.5744\angle{59.1°}\ 12.2\angle{-49.1°} + 83.51\angle{0°}\\ &= 90 + j1.15 = 90\angle{0.732°}\ \text{V}\end{aligned} \tag{5.71}$$

SP5.4.1 Calculate $\widetilde{I}_{as}$ for $T_L = 5$ N m at Operating Point 2.

SP5.4.2 Draw the phasor diagram for Operating Point 2 of Figure 5.17.

SP5.4.3 What changes are necessary for a three-phase induction motor? $\left[L_{ms}\right.$ and T_e are multiplied by $\left.\frac{3}{2}\right]$

5.5 Problems

5.1 The parameters of a permanent-magnet dc machine are $r_a = 6\ \Omega$ and $k_v = 2\times10^{-2}$ V s/rad and V_a can be varied from zero to 10 volts. The device is to be operated in the constant-torque mode with $T_e = 4\times10^{-3}$ N m. (a) Determine V_a for $\omega_r = 0$. (b) Determine maximum ω_r range of the constant-torque mode of operation that is, maximum ω_r with $T_e = 4\times10^{-3}$ N m and $V_a = 10$ V.

5.2 Sketch Figure 5.2 for generator action; i.e. i_{S1} and i_{D2} equal to zero.

5.3 The parameters of a two-pole three-phase permanent magnet ac machine are $r_s = 4\ \Omega$, $\lambda_m'^r = 0.07$ V s/rad, $L_{ls} = 1$ mH, $L_{ms} = 9$ mH. The applied voltages are $V_{as} = \sqrt{2}\,14.1\cos\theta_r$ and $V_{bs} = \sqrt{2}\,14.1\sin\theta_r$ where $\theta_r = 200t$. Determine (a) $\tilde{I}_{as}$, (b) T_e, and (c) show the phasor diagram and locate the poles.

5.4 Consider the machine given in Section 5.3. It is stated that the rotor speed is $\omega_r = 120.1$ rad/s where 0.315 N m can no longer be maintained with $\phi_v = 0$ and $V_s = 11.25$ V. Verify this statement.

5.5 Plot v_{as}, v_{bs}, v_{cs}, v_{ab}, and v_{np} for the 180° continuous-current inverter for $0 < \theta_r < 2\pi$ with the transistor switching shown in Figure 5.7. Assume the inverter voltage v_i is constant.

5.6 Draw the phasor diagram and locate the poles for Problem 5.4 above.

5.7 Consider the two-phase 5-hp induction machine given in Section 5.4. Establish (a) the slip frequency ω_e-ω_r and (b) $\tilde{I}_{as}$ for $T_e = 5$ N m and $I_{ds}^e = 12.99$ A.

References

1 Krause, P. C., Wasynczuk, O., Pekarek, S. D., and O'Connell, T. C. (2020). *Electromechanical Motion Devices*, 3e, IEEE Press/Wiley.

2 Krause, P. C., Wasynczuk, O., Sudhoff, S. D., and Pekarek, S. D. (2013). *Analysis of Electric Machinery and Drive Systems*, 3e, IEEE Press/Wiley.

3 Krause, P. C., Wasynczuk, O., and Sudhoff, S. D. (1994). *Analysis of Electric Machinery*, 1e, IEEE Press.

4 Trzynadlowski, A. M. (1994). *The Field Orientation Principle in Control of Induction Motors*, Kluwer Academic Publishers.

5 Blaschke, F. (1974). Das verfahren der feldorientierung zur regelung der drehfeld-maschine, PhD thesis. TU Braunschweig.

6

Power Systems

6.1 Introduction

Although power system engineering dominated the power area in the first half of the twentieth century, electric drives, power electronics, and machine design have emerged as important areas of power engineering. It is understandable that power system analysis having evolved first, established conventions and analysis techniques appropriate for power systems and not necessarily convenient for the areas of power engineering that came later. For example, the power system engineer prefers to consider positive current out of the machine since they work primarily with power generation, while the electric drive's engineer deals primarily with motors and the power system's engineering convention is not the most convenient. In this chapter, several of the analytical aspects of power system engineering are set forth. Although the presentation may be a bit disjointed, the power system engineer is faced with a host of challenges and our purpose is to make the reader aware of some of these challenges without becoming too involved.

6.2 Three-phase Transformer Connections

Single-phase transmission exists in urban distribution systems and sparsely populated rural areas. Three-phase systems have traditionally been associated with electric power generation and high-voltage transmission. Therefore, three-phase transformer connections warrant a brief consideration in an introductory treatment of power system engineering.

In a two-phase system, transformers are generally arranged as two separate single-phase transformers without interconnection; however, in a three-phase system, the phase transformers are generally interconnected and, although there are numerous variations, there are essentially only three basic transformer configurations. We will touch on each of these. Three-phase transformers may

Introduction to the Analysis of Electromechanical Systems, First Edition.
Paul C. Krause, Oleg Wasynczuk, and Timothy O'Connell.

be three separately constructed single-phase transformers or three single-phase transformers wound on a common iron core. We will deal only with three separate single-phase transformers.

6.2.1 Wye–Wye Connection

The so-called wye–wye (Y–Y) connection is illustrated in Figure 6.1. The transformers are identical; hence $r_{a1} = r_{b1} = r_{c1}$, each with N_1 turns and $r_{a2} = r_{b2} = r_{c2}$, each with N_2 turns. Here, we have changed notation to accommodate the three-phase variables. Rather than denoting the windings as winding 1 and winding 2, as in previous sections, the notation of $a1$ and $a2$ winding, respectively, is used for the a-phase transformer, $b1$ and $b2$ winding for the b-phase transformer, and $c1$ and $c2$ winding for the c-phase transformer. In Figure 6.1a, the three single-phase transformers are arranged to show the transformer coupling in each phase; however, the circuit diagram shown in Figure 6.1b is generally used since it most readily depicts the type of connection being used. Although the Y–Y connection is not used too often in power systems, it is a good place to start.

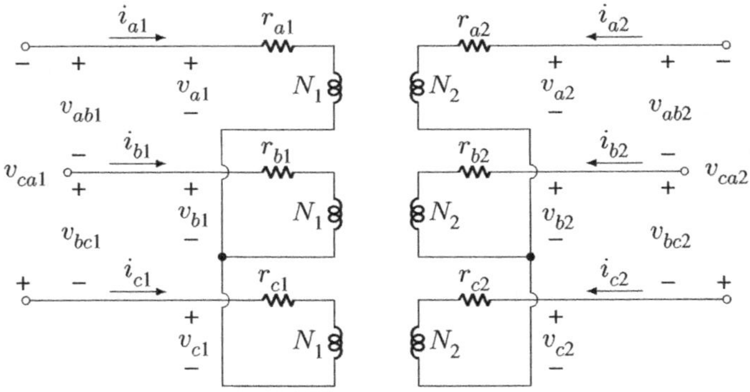

(a)

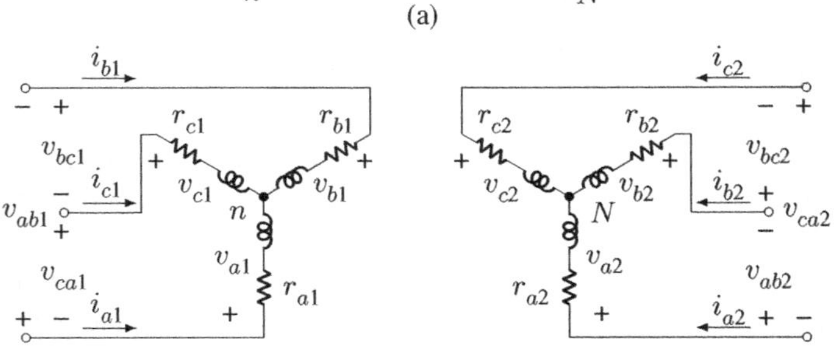

(b)

Figure 6.1 A Y–Y three-phase transformer connection.

Note that only three conductors enter and leave the transformers. This does not seem correct. For a single-phase system we would need two conductors, and for a two-phase system we would need four. Why are only three required for a three-phase system? Well, we know that a balanced three-phase set of currents consists of, by definition, equal-amplitude sinusoidal currents displaced by 120°. The sum of these currents is zero; hence, if the system is balanced, $i_{a1} + i_{b1} + i_{c1} = 0$, the current entering one phase of the 1-winding returns through the other phases of the 1-windings. Actually, a fourth wire is not needed regardless if the system is balanced or unbalanced. We are now starting to see the economic advantage of transmitting three-phase electric power. There is also another advantage of two- or three-phase electromechanical devices over their single-phase counterparts. As we mentioned earlier, the instantaneous power or torque in a single-phase system pulsates about an average value; however, in a multiphase system, the instantaneous power or torque is constant for balanced steady-state conditions and the power system operator strives to keep the system balanced.

In a Y–Y connection, the assigned negative-potential sides of the windings are all connected to form what is called the *neutral point*; *n* for 1-windings in Figure 6.1 and *N* for the 2-windings. Either or both neutral points may be grounded or left to "float." The voltages across each transformer winding, v_{a1}, v_{a2}, v_{b1}, etc. are referred to as phase voltages or line-to-neutral voltages, while the voltages between two of the three terminals of the 1-windings are the line-to-line voltages. The individual transformer currents i_{a1}, i_{a2}, i_{b1}, etc. are the phase or line currents. The line-to-neutral and the line-to-line voltages of the 1-windings are related as

$$v_{ab1} = v_{a1} - v_{b1} \tag{6.1}$$

$$v_{bc1} = v_{b1} - v_{c1} \tag{6.2}$$

$$v_{ca1} = v_{c1} - v_{a1} \tag{6.3}$$

If the system is balanced, then we can express the steady-state voltages of the 1-windings for an *abc* sequence as

$$V_{a1} = \sqrt{2}V_{s1} \cos \omega_e t \tag{6.4}$$

$$V_{b1} = \sqrt{2}V_{s1} \cos\left(\omega_e t - \frac{2\pi}{3}\right) \tag{6.5}$$

$$V_{c1} = \sqrt{2}V_{s1} \cos\left(\omega_e t + \frac{2\pi}{3}\right) \tag{6.6}$$

where the uppercase letters are used to denote steady-state conditions. In phasor form,

$$\tilde{V}_{a1} = V_{s1}\underline{/0^\circ} \tag{6.7}$$

$$\tilde{V}_{b1} = V_{s1}\angle -120^\circ \tag{6.8}$$

$$\tilde{V}_{c1} = V_{s1}\angle 120^\circ \tag{6.9}$$

From (6.1–6.3), the line-to-line voltage may be expressed as

$$\begin{aligned}\tilde{V}_{ab1} &= V_{s1}\angle 0^\circ - V_{s1}\angle -120^\circ \\ &= \sqrt{3}V_{s1}\angle 30^\circ\end{aligned} \tag{6.10}$$

$$\begin{aligned}\tilde{V}_{bc1} &= V_{s1}\angle -120^\circ - V_{s1}\angle 120^\circ \\ &= \sqrt{3}V_{s1}\angle -90^\circ\end{aligned} \tag{6.11}$$

$$\begin{aligned}\tilde{V}_{ca1} &= V_{s1}\angle 120^\circ - V_{s1}\angle 0^\circ \\ &= \sqrt{3}V_{s1}\angle 150^\circ\end{aligned} \tag{6.12}$$

Hence, the line-to-line voltages form a balanced three-phase set which are $\sqrt{3}$ times the magnitude of the line-to-neutral voltages and shifted in phase by 30°.

For balanced steady-state conditions, the operation of each phase may be considered separately. In fact, we need to consider only one phase since, once we have determined the variables associated with one phase of the transformers, we can express the other phase variables by shifting the phase by either 120° and −120°. The above voltage equations apply to the 2 winding, we only need to change the 1 subscript to 2.

6.2.2 Delta–Delta Connection

The delta–delta ($\Delta - \Delta$) connection is illustrated in Figure 6.2. In this type of connection, the line-to-line voltages are the voltages across the individual transformers, that is, $v_{a1} = v_{ab1}, v_{a2} = v_{ab2}, v_{b1} = v_{bc1}$, etc. There is no neutral connection. The line currents are made up of the currents from two of the transformers. In particular,

$$i_{ac1} = i_{a1} - i_{c1} \tag{6.13}$$

$$i_{ba1} = i_{b1} - i_{a1} \tag{6.14}$$

$$i_{cb1} = i_{c1} - i_{b1} \tag{6.15}$$

If, for example, the transformer currents form a balanced *abc* sequence, then

$$\tilde{I}_{a1} = I_{s1}\angle 0^\circ \tag{6.16}$$

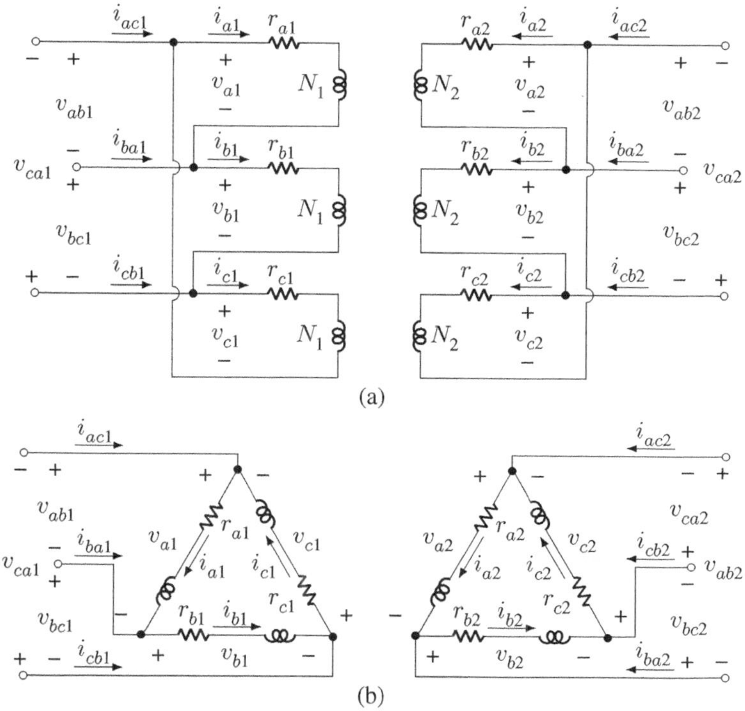

Figure 6.2 A Δ - Δ three-phase transformer connection.

$$\tilde{I}_{b1} = I_{s1}\underline{/-120^\circ} \tag{6.17}$$

$$\tilde{I}_{c1} = I_{s1}\underline{/120^\circ} \tag{6.18}$$

The line currents become

$$\begin{aligned}\tilde{I}_{ac1} &= I_{s1}\underline{/0^\circ} - I_{s1}\underline{/120^\circ} \\ &= \sqrt{3}I_{s1}\underline{/-30^\circ}\end{aligned} \tag{6.19}$$

Thus, $\tilde{I}_{ac1}$ is $\sqrt{3}$ times the amplitude of $\tilde{I}_{a1}$ and shifted 30° back in phase from it. Similarly, $\tilde{I}_{ba1}$ is shifted back 30° from $\tilde{I}_{b1}$, and $\tilde{I}_{cb1}$ back 30° from $\tilde{I}_{c1}$. For balanced conditions, a phase current in a $Y - Y$ connection is $\sqrt{3}$ times the phase current in a $\Delta - \Delta$ connection for the same line current.

6.2.3 Wye–Delta or Delta–Wye Connection

The final three-phase transformer connection that we will consider is the wye–delta ($Y - \Delta$) or delta–wye ($\Delta - Y$) connection. The $\Delta - Y$ is generally used to

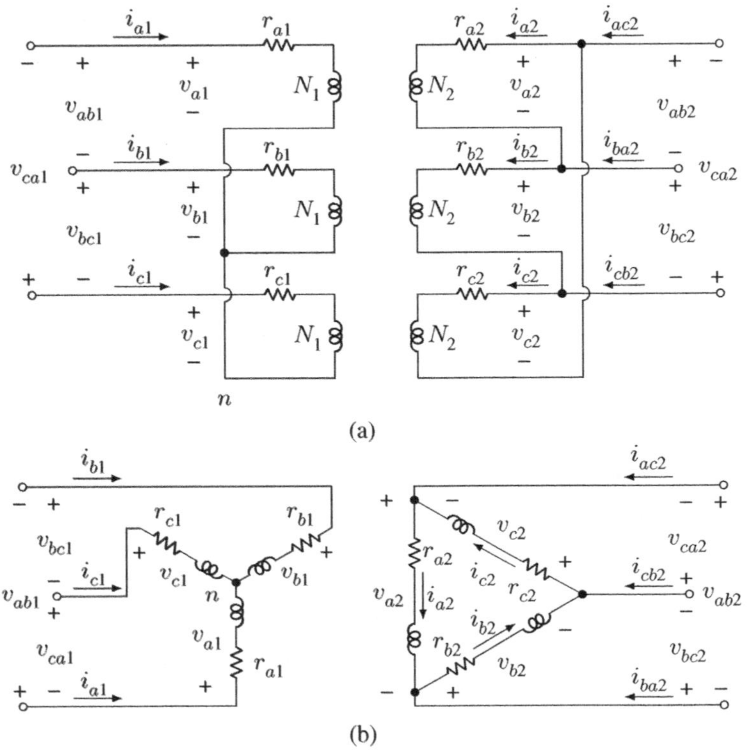

Figure 6.3 A $Y-\Delta$ three-phase transformer connection.

"step up" the voltage thus the $Y-\Delta$ is used for "stepping down" the voltage. The $Y-\Delta$ connection is illustrated in Figure 6.3.

6.2.4 Ideal Transformers

Since in electromechanical motion devices the magnetizing flux must traverse the air gap, the leakage inductances and the magnetizing current must be considered when analyzing these devices. However, in the analysis of a power system, the system transformers are often assumed to be ideal. In this case, the winding resistance, leakage inductance, and magnetizing current are all neglected and the voltages (currents) of the transformer windings differ only by the turns ratio. The relations between 1- and 2-winding variables for the $Y-Y$, $\Delta-\Delta$, and $Y-\Delta$ connections of ideal transformers are shown in Figure 6.4.

Example 6A ***Equivalent wye- and delta-connected impedances***
Determine the equivalent delta and wye impedances.

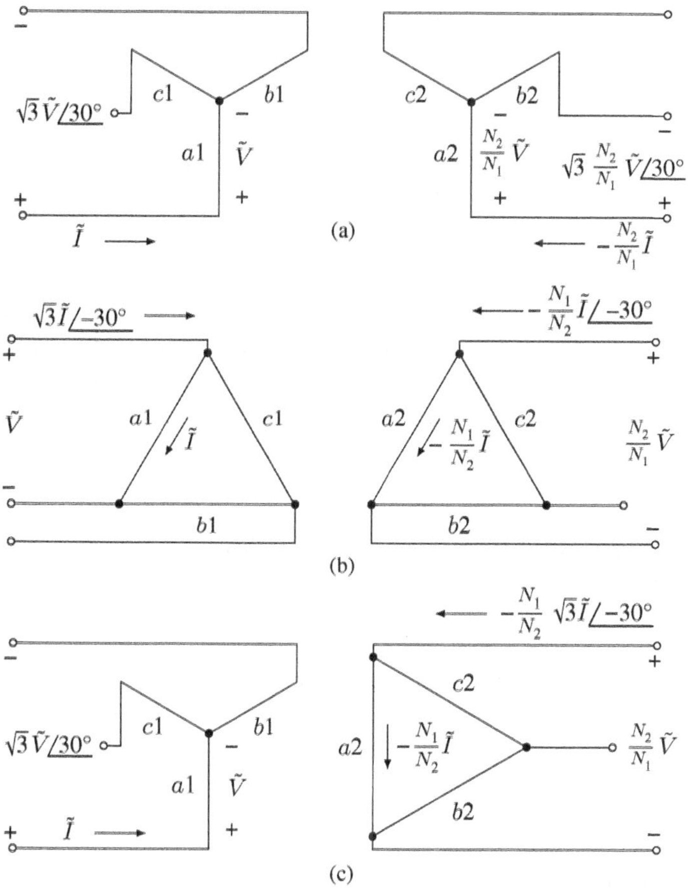

Figure 6.4 Three-phase ideal transformers for *abc* sequence: (a) Y – Y connection; (b) Δ – Δ connection; (c) Y – Δ connection.

From Figure 6A.1

$$Z_a + Z_b = \frac{Z_{ab}(Z_{ca} + Z_{bc})}{Z_{ab} + Z_{bc} + Z_{ca}} \tag{6A.1}$$

$$Z_b + Z_c = \frac{Z_{bc}(Z_{ab} + Z_{ca})}{Z_{ab} + Z_{bc} + Z_{ca}} \tag{6A.2}$$

$$Z_c + Z_a = \frac{Z_{ca}(Z_{bc} + Z_{ab})}{Z_{ab} + Z_{bc} + Z_{ca}} \tag{6A.3}$$

Show that if Z_{ab}, Z_{bc}, and Z_{ca} are each equal to Z_Δ and Z_a, Z_b, and Z_c equal to Z_Y then

$$Z_\Delta = 3Z_Y \tag{6A.4}$$

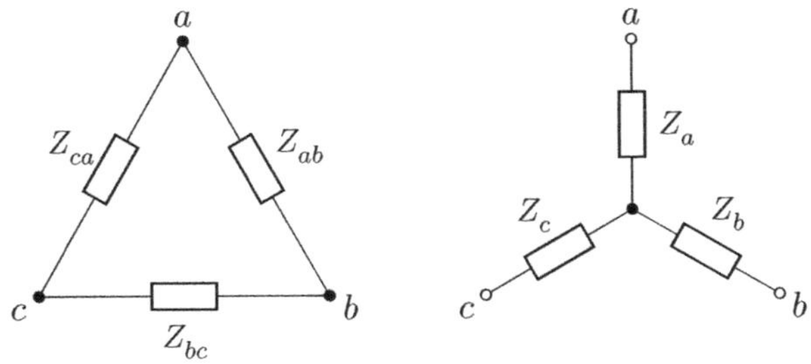

Figure 6A.1 Delta- and wye-connected impedances.

SP6.2.1 Write the expression for the steady-state line-to-line voltage V_{ab} of a Y−Y connected three-phase system for a balanced set of (a) *abc* sequence voltages and (b) *acb* sequence voltages. Assume that the steady-state line-to-neutral voltage of phase *a* is $V_{a1} = \sqrt{2}V_s\cos(\omega_e t - 30°)$. [(a) $V_{ab} = \sqrt{6}V_s\cos\omega_e t$; (b) $V_{ab} = \sqrt{6}V_s\cos(\omega_e t - 60°)$]

SP6.2.2 Refer to Figure 6.2. Calculate I_{ac1} for an *acb* sequence. Assume $I_{a1} = I_{s1}\underline{/0°}$. [$\tilde{I}_{ac1} = \sqrt{3}I_{s1}\underline{/30°}$]

SP6.2.3 In Figure 6.1, $V_{ab1} = \sqrt{6}\cos\omega_e t$ and $I_{a1} = \sqrt{2}\cos(\omega_e t - 30°)$. Assume balanced operation with an *abc* sequence. Calculate the total power flow into the Y−Y transformer. [3 W]

SP6.2.4 Show that the power supplied to a balanced wye and an equivalent delta connected resistors is equal.

6.3 Synchronous Generator

Although the synchronous machine can be operated as either a generator or a motor, it is generally operated as a generator. In the steady state, the rotor rotates in synchronism with the electrical angular velocity corresponding to that of the system to which it is connected; 60 Hz in North and South America, 50 Hz in most of Europe, Asia, and Africa. In Chapter 5, we used the permanent-magnet alternating current (ac) machine, which is also a synchronous machine, as the motor in an electric drive. In that case, the stator windings of the permanent-magnet ac machine were supplied from an inverter that was controlled so that the frequency of the applied stator voltages was made equal to the electrical angular velocity of the rotor. In other words, the isolated machine is always operating in synchronism with the supply voltages. That is, the motor slows when it is loaded and the frequency of the stator voltages is decreased in unison with the decrease in rotor speed.

The synchronous machine operating in a power grid is a much different situation. In this case, the power grid operates at essentially a constant frequency and the synchronous machine must operate in synchronism with the frequency of the power grid. The synchronous generator "falling out" of synchronism with the power system during a disturbance is a huge concern for power system operators. Therein lies an area of system stability that presents many engineering challenges. Our purpose here is to introduce the synchronous generator and in doing so we will draw on material in Section 3.3.

In the case of isolated operation of a small electrical power system with one or two generators, the frequency (electrical angular velocity) of the system may vary somewhat depending on the load and the power generated. Some aircraft, ships, or microgrids are examples; however, we will not consider this mode of operation.

A three-phase round-rotor synchronous machine is shown in Figure 6.5. We have replaced the permanent-magnet rotor of the permanent-magnet ac machine with a field winding; the *fd*-winding. The voltage equations are

$$v_{as} = r_s i_{as} + \frac{d\lambda_{as}}{dt} \tag{6.20}$$

$$v_{bs} = r_s i_{bs} + \frac{d\lambda_{bs}}{dt} \tag{6.21}$$

$$v_{cs} = r_s i_{cs} + \frac{d\lambda_{cs}}{dt} \tag{6.22}$$

$$v_{fd} = r_f i_{fd} + \frac{d\lambda_{fd}}{dt} \tag{6.23}$$

Equations (6.20)–(6.22) are identical in form to the three-phase permanent-magnet ac machine. Therefore, v_{qs}^r and v_{ds}^r will be identical in form as (3.33) and

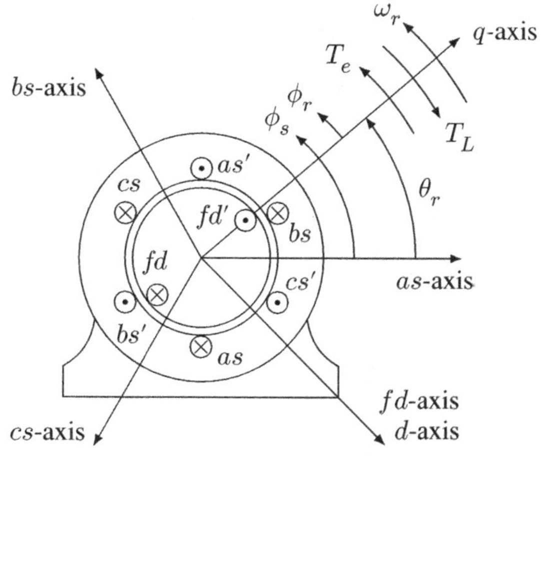

Figure 6.5 Three-phase round-rotor synchronous machine.

(3.34). The voltage equation for v_{0s} is omitted since we are dealing only with balanced conditions. The *fd*-winding is physically in the rotor frame of reference and therefore will not require a change of variables. In fact, the steady-state voltage equations are similar to those of the permanent-magnet ac machine, differing only in that the synchronous machine (generator) has a field winding rather than a permanent magnet. If the focus is on the analysis of steady-state operation, the reader may wish to deemphasize the material up to Subsection 6.3.3.

To derive the voltage equations valid for dynamic operation, it is necessary to consider the flux linkage equations; in particular

$$\begin{bmatrix} \lambda_{as} \\ \lambda_{bs} \\ \lambda_{cs} \\ \lambda_{fd} \end{bmatrix} = \begin{bmatrix} L_{ls} + L_{ms} & -\frac{1}{2}L_{ms} & -\frac{1}{2}L_{ms} & L_{sfd}\sin\theta_r \\ -\frac{1}{2}L_{ms} & L_{ls} + L_{ms} & -\frac{1}{2}L_{ms} & L_{sfd}\sin(\theta_r - 120^\circ) \\ -\frac{1}{2}L_{ms} & -\frac{1}{2}L_{ms} & L_{ls} + L_{ms} & L_{sfd}\sin(\theta_r + 120^\circ) \\ L_{sfd}\sin\theta_r & L_{sfd}\sin(\theta_r - 120^\circ) & L_{sfd}\sin(\theta_r + 120^\circ) & L_{lfd} + L_{mfd} \end{bmatrix} \begin{bmatrix} i_{as} \\ i_{bs} \\ i_{cs} \\ i_{fd} \end{bmatrix} \tag{6.24}$$

This may be written

$$\begin{bmatrix} \boldsymbol{\lambda}_{abcs} \\ \lambda_{fd} \end{bmatrix} = \begin{bmatrix} \mathbf{L}_s & \mathbf{L}_{sr} \\ (\mathbf{L}_{sr})^T & L_{fdfd} \end{bmatrix} \begin{bmatrix} \mathbf{i}_{abcs} \\ i_{fd} \end{bmatrix} \tag{6.25}$$

where

$$\mathbf{L}_s = \begin{bmatrix} L_{ls} + L_{ms} & -\frac{1}{2}L_{ms} & -\frac{1}{2}L_{ms} \\ -\frac{1}{2}L_{ms} & L_{ls} + L_{ms} & -\frac{1}{2}L_{ms} \\ -\frac{1}{2}L_{ms} & -\frac{1}{2}L_{ms} & L_{ls} + L_{ms} \end{bmatrix} \tag{6.26}$$

$$\mathbf{L}_{sr} = \begin{bmatrix} L_{sfd}\sin\theta_r \\ L_{sfd}\sin(\theta_r - 120^\circ) \\ L_{sfd}\sin(\theta_r + 120^\circ) \end{bmatrix} \tag{6.27}$$

$$L_{fdfd} = L_{lfd} + L_{mfd} \tag{6.28}$$

Since we are assuming a magnetically linear system, the inductances may be expressed

$$L_{ms} = \frac{N_s^2}{\mathcal{R}_m} \tag{6.29}$$

$$L_{mfd} = \frac{N_f^2}{\mathcal{R}_m} = \frac{N_f^2}{N_s^2} L_{ms} \tag{6.30}$$

$$L_{sfd} = \frac{N_s N_f}{\mathcal{R}_m} = \frac{N_f}{N_s} L_{ms} \tag{6.31}$$

where N_s (N_f) is the turns of the stator windings (*fd*-winding). Now, from (6.25)

$$\begin{aligned}\boldsymbol{\lambda}^r_{qd0s} &= \mathbf{K}^r_s\boldsymbol{\lambda}_{abcs} + \mathbf{K}^r_s\mathbf{L}_{sr}i^r_{fd} \\ &= \mathbf{K}^r_s\mathbf{L}_s(\mathbf{K}^r_s)^{-1}\mathbf{i}^r_{qd0s} + \mathbf{K}^r_s\mathbf{L}_{sr}i^r_{fd}\end{aligned} \tag{6.32}$$

Now, since $i_{as} + i_{bs} + i_{cs} = 0$, (6.32) may be written

$$\begin{bmatrix}\lambda^r_{qs}\\ \lambda^r_{ds}\\ \lambda_{0s}\end{bmatrix} = \begin{bmatrix} L_{ls}+\frac{3}{2}L_{ms} & 0 & 0\\ 0 & L_{ls}+\frac{3}{2}L_{ms} & 0\\ 0 & 0 & L_{ls}\end{bmatrix}\begin{bmatrix}i^r_{qs}\\ i^r_{ds}\\ i_{0s}\end{bmatrix} + L_{sfd}\begin{bmatrix}0\\1\\0\end{bmatrix}i^r_{fd} \tag{6.33}$$

In (6.32) and (6.33), we have added an r superscript to i_{fd} to emphasize that it is in the rotor reference frame. Now, in preparation for an equivalent circuit with $\frac{3}{2}L_{ms}$ as the mutual inductance between stator and rotor circuits, let us express (6.33) as

$$\lambda^r_{qs} = L_{ls}i^r_{qs} + \frac{3}{2}L_{ms}i^r_{qs} \tag{6.34}$$

$$\lambda^r_{ds} = L_{ls}i^r_{ds} + \frac{3}{2}L_{ms}(i^r_{ds} + i'^r_{fd}) \tag{6.35}$$

$$\lambda_{0s} = L_{ls}i_{0s} \tag{6.36}$$

where

$$i'^r_{fd} = \frac{2}{3}\frac{N_f}{N_s}i^r_{fd} \tag{6.37}$$

Comparing the last term of (6.33) and the last term of (6.35), we see that

$$\begin{aligned}L_{sfd}i^r_{fd} &= \frac{N_f}{N_s}L_{ms}\frac{N_s}{N_f}\frac{3}{2}i'^r_{fd} \\ &= \frac{3}{2}L_{ms}i'^r_{fd}\end{aligned} \tag{5.38}$$

If we let

$$v'^r_{fd} = \frac{N_s}{N_f}v^r_{fd} \tag{6.39}$$

$$\lambda'^r_{fd} = \frac{N_s}{N_f}\lambda^r_{fd} \tag{6.40}$$

and after considerable algebraic manipulation, the turns-ratio referred field-voltage equation becomes

$$v'^r_{fd} = r'_{fd}i'^r_{fd} + p\lambda'^r_{fd} \tag{6.41}$$

where

$$\lambda'^r_{fd} = L'_{lfd} i'^r_{fd} + \frac{3}{2} L_{ms} (i^r_{ds} + i'^r_{fd}) \tag{6.42}$$

$$r'_{fd} = \frac{3}{2} \left(\frac{N_s}{N_f} \right)^2 r_{fd} \tag{6.43}$$

$$L'_{lfd} = \frac{3}{2} \left(\frac{N_s}{N_f} \right)^2 L_{lfd} \tag{6.44}$$

Care must be taken since

$$v^r_{fd} i^r_{fd} \neq v'^r_{fd} i'^r_{fd} \tag{6.45}$$

however, (6.37) allows a convenient equivalent circuit that results by combining (6.41) with (3.33) and (3.34) which are repeated here

$$v^r_{qs} = r_s i^r_{qs} + \omega_r \lambda^r_{ds} + p\lambda^r_{qs} \tag{6.46}$$

$$v^r_{ds} = r_s i^r_{ds} - \omega_r \lambda^r_{qs} + p\lambda^r_{ds} \tag{6.47}$$

where, in the case of the three-phase machine

$$\lambda^r_{qs} = L_{ls} i^r_{qs} + \frac{3}{2} L_{ms} i^r_{qs} \tag{6.48}$$

$$\lambda^r_{ds} = L_{ls} i^r_{ds} + \frac{3}{2} L_{ms} (i^r_{ds} + i'^r_{fd}) \tag{6.49}$$

The equivalent circuit suggested by (6.41), (6.42), and (6.46)–(6.49) is shown in Figure 6.6. The zero variables are not shown since we will not consider unbalanced conditions.

6.3.1 Damper Windings

We need to take a moment to mention a characteristic of the synchronous machine that, in the past, caused some concern and to indicate how it was remedied. The device shown in Figure 6.5 provides very little damping to rotor oscillations following system disturbances that can cause undesirable fluctuations in power flow. Damping of these rotor oscillations is achieved by currents flowing in short-circuited rotor bars that provide a damping torque by induction motor action. That is, currents are introduced in these "windings" only when the rotor is not revolving in unison with the rotating air-gap magnetomotive force (mmf) established by the stator currents. Since these windings do not play a role during balanced steady-state operation, we will not consider them in this derivation.

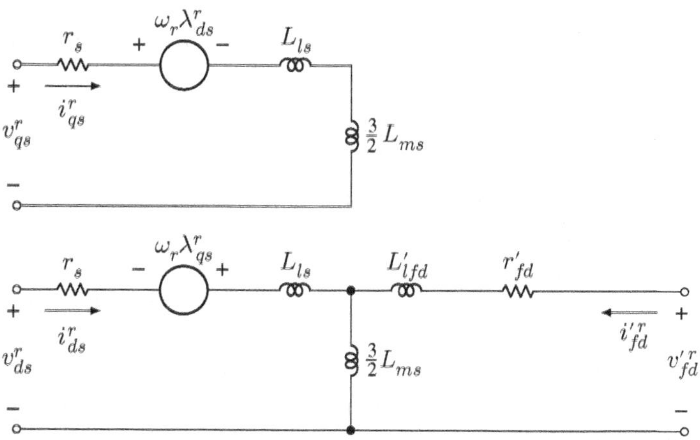

Figure 6.6 The q^r_s and d^r_s equivalent circuit for a three-phase round rotor synchronous machine.

6.3.2 Torque

Recall in Section 2.3 we expressed the torque by relating the energy in the coupling field, the energy entering the coupling field from the electrical system and the energy entering from the mechanical system. There are other methods of deriving an expression for torque. For example, once the variables are expressed in the rotor reference frame, torque can be conveniently expressed using a "power balance" approach. For balanced conditions, the electrical power of the generator can be expressed

$$P_e = \frac{3}{2}(v^r_{qs}i^r_{qs} + v^r_{ds}i^r_{ds}) \tag{6.50}$$

The $\frac{3}{2}$ factor comes about due to the transformation of a three-phase system to a two-phase system. The mechanical power can be expressed

$$P_m = T_e\left(\frac{2}{P}\right)\omega_r \tag{6.51}$$

where T_e is positive for motor action. Substituting (6.46) and (6.47) for v^r_{qs} and v^r_{ds}, into (6.50) and then the coefficients of ω_r is the electromagnetic torque which is positive for motor action.

$$T_e = \left(\frac{3}{2}\right)\left(\frac{P}{2}\right)(\lambda^r_{ds}i^r_{qs} - \lambda^r_{qs}i^r_{ds}) \tag{6.52}$$

The torque and speed may be related as given by (2.114) which is repeated here

$$T_e = J\left(\frac{2}{P}\right)\frac{d\omega_r}{dt} + B_m\left(\frac{2}{P}\right)\omega_r + T_L \tag{6.53}$$

where P is the number of poles, J is the inertia in kg m^2, B_m is the damping coefficient in N m s/rad due to friction and windage losses, and T_L is positive for a load torque.

6.3.3 Steady-state Operation and Rotor Angle

The analysis of steady-state operation of the synchronous generator is very similar to that carried out in the case of the permanent-magnet ac machine in Section 3.3. Therein, it was convenient to fix the time-zero position of the q axis on the real axis and for conventional brushless direct current (dc) motor operation, the firing of the inverter is controlled to fix the time-zero position of the as-phase voltage (phasor $\widetilde{V}_{as}$) on the real axis, coinciding with the q axis as shown in Figure 3C.1. However, as discussed in Chapter 5, we can shift the applied voltages by controlling the firing of the inverter to maximize torque. The important difference in the operation of the permanent-magnet ac machine in an electric drive system and that of a synchronous generator connected to a large power system is that the large power system is generally not influenced by the mode of operation of the smaller generator. That is, the power system voltages remain fixed in frequency, phase, and magnitude. In this case, the power system is referred to as an "infinite bus" since its mechanical and electrical momentums are not influenced by the operation of one synchronous generator. Therefore, in the analysis of the synchronous generator, the phase of the system voltages is generally assumed to remain fixed and the time-zero position is selected so that, the system voltage, $\widetilde{V}_{as}$ is fixed on the real axis. The q and d axes, which are fixed in the rotor, shift in position relative to the synchronous speed of the electric power system (phase of $\widetilde{V}_{as}$) as a result of a change in shaft torque. Therefore, we are not surprised that there could be a problem keeping the generator operating in synchronism with the power grid. This is of particular concern during a disturbance such as a fault near the generator resulting in a disruption of power flow from the generator to the system. In this case, the torque input to the generator would accelerate the generator above synchronous speed. If the fault isn't cleared and generator operation rapidly restored, its power delivery capability to the system is lost. Clearly, this was not of concern in the case of the permanent-magnet ac machine in an inverter drive since the source frequency is controlled to be the same as the electrical angular velocity of the rotor.

During steady-state operation of the synchronous generator, the q- and d-voltage equations, (6.46), (6.47), and (6.41), respectively, become

$$V_{qs}^r = r_s I_{qs} + \omega_r \lambda_{ds}^r \tag{6.54}$$

$$V_{ds}^r = r_s I_{ds}^r - \omega_r \lambda_{qs}^r \tag{6.55}$$

$$V'^r_{fd} = r'_{fd} I'^r_{fd} \tag{6.56}$$

The rotor angle, which is the displacement of the q axis relative to $\widetilde{V}_{as}$, is defined as

$$\delta = \theta_r - \theta_{esv}$$
$$= \int_0^t [\omega_r(\xi) - \omega_e(\xi)]d\xi + \theta_r(0) - \theta_{esv}(0) \tag{6.57}$$

In the steady state $\omega_r = \omega_e$ and

$$\delta = \theta_r(0) - \theta_{esv}(0) \tag{6.58}$$

Let us now consider a balanced three-phase system with

$$F_{as} = \sqrt{2}\,F_s \cos\theta_{esf} \tag{6.59}$$

For steady-state conditions θ_{esf} is $\omega_e t + \theta_{esf}(0)$ and the q_s^r- and d_s^r- variables are

$$F_{qs}^r = \sqrt{2}F_s \cos[\theta_{esf}(0) - \theta_r(0)] \tag{6.60}$$

$$F_{ds}^r = -\sqrt{2}F_s \sin[\theta_{esf}(0) - \theta_r(0)] \tag{6.61}$$

If we solve (6.58) for $\theta_r(0)$ and substitute the result into (6.60) and (6.61), we will obtain

$$F_{qs}^r = \sqrt{2}F_s \cos[\theta_{esf}(0) - \theta_{esv}(0) - \delta] \tag{6.62}$$

$$F_{ds}^r = -\sqrt{2}F_s \sin[\theta_{esf}(0) - \theta_{esv}(0) - \delta] \tag{6.63}$$

As we have mentioned, when the synchronous generator is connected to a large power system, the system voltage is unchanged in magnitude and phase by the synchronous generator. We generally select time zero when the system voltage at the point of connection is maximum, thus $\theta_{esv}(0) = 0$. Therefore, $\theta_{esv}(0)$ is zero in (6.58) and if (6.59–6.61) are used to express voltages, $\theta_{esf}(0)$ in these equations is zero since in this case it is $\theta_{esv}(0)$.

Now, from (6.59)

$$\widetilde{F}_{as} = F_s e^{j\theta_{esf}(0)} \tag{6.64}$$

Multiplying both sides of (6.64) by $e^{-j\delta}$ yields

$$\widetilde{F}_{as} e^{-j\delta} = F_s \cos[\theta_{esf}(0) - \delta] + jF_s \sin[\theta_{esf}(0) - \delta] \tag{6.65}$$

where $\theta_{esv}(0)$ in (6.62) and (6.63) has been set equal to zero. Comparing (6.65) with (6.62) and (6.63) we see that, since $\theta_{esv}(0)$ is zero, $\widetilde{F}_{as}$ may be expressed

$$\sqrt{2}\widetilde{F}_{as} = (F_{qs}^r - jF_{ds}^r)e^{j\delta} \tag{6.66}$$

Following the same procedure as in the case of a round-rotor permanent-magnet ac machine, we obtain

$$\widetilde{V}_{as} = (r_s + j\omega_e L_{ss})\widetilde{I}_{as} + \frac{1}{\sqrt{2}}\omega_e\left(\frac{3}{2}\right)L_{ms}I'^{r}_{fd}e^{j\delta} \tag{6.67}$$

If we compare the last term of (3.60) and (6.67), we see that λ'^{r}_{m} is analogous to $\frac{3}{2}L_{ms}I'^{r}_{fd}$. Also in (3.59), the last term, which we called $\widetilde{E}_a$, was on the q axis, fixed at zero degrees whereas the last term in (6.67), which we are also going to call $\widetilde{E}_a$, is also along the q axis but positioned at the angle δ. That is

$$\widetilde{E}_a = \frac{1}{\sqrt{2}}\,\omega_e\left(\frac{3}{2}\right)L_{ms}I'^{r}_{fd}e^{j\delta} \tag{6.68}$$

In the case of the brushless dc drive, we have the capability to change instantaneously the phase relation between the permanent magnet of the rotor and the applied voltages by controlling the phase (shifting) of the applied voltages. The permanent magnet is fixed in the d axis and the q axis of the rotor and thus $\widetilde{E}_a$ are fixed 90° ahead (counterclockwise) of the d axis. In the case of the synchronous generator connected to an electric power grid, however, the stator voltages are fixed in phase and magnitude by the power grid. In this case, the system voltages are the fixed reference and $\widetilde{V}_{as}$ is selected at zero degrees and the q axis and thus $\widetilde{E}_a$ shifts relative to $\widetilde{V}_{as}$ depending on the operating conditions (loading) of the generator.

The steady-state torque may be expressed from (6.48), (6.49), and (6.52) as

$$T_e = \left(\frac{P}{2}\right)\left(\frac{3}{2}\right)L_{ms}I^{r}_{qs}I'^{r}_{fd} \tag{6.69}$$

This relationship can also be obtained by realizing from (6.67) that λ'^{r}_{m} is analogous to $\frac{3}{2}L_{ms}I'^{r}_{fd}$; thus, (6.69) can be justified for steady-state operation from (3.63).

It is common to neglect r_s in (6.54) and (6.55), which is a good approximation in the case of a 60 or 50-Hz power system since the inductive reactance dominates the machine impedance. If we then solve (6.55) for I^{r}_{qs} and substitute it into (6.69), we can express the steady-state torque, with r_s neglected, as

$$T_e = -\left(\frac{P}{2}\right)\left(\frac{3}{2}\right)\left(\frac{1}{\omega_e}\right)\left(\frac{1}{X_{ss}}\right)E'_x\sqrt{2}V_s\sin\delta \tag{6.70}$$

where $\omega_r = \omega_e$, and E'_x, the excitation voltage, is

$$E'_x = \omega_e\left(\frac{3}{2}\right)L_{ms}I'^{r}_{fd} = \sqrt{2}\,|\widetilde{E}_a| \tag{6.71}$$

From (6.70) it is clear that stable steady-state operation occurs for $-\frac{\pi}{2} < \delta < \frac{\pi}{2}$. Where motor operation occurs for $-\frac{\pi}{2} < \delta < 0$ and generator operation for $0 < \delta < \frac{\pi}{2}$.

Let us take a few minutes to discuss the operation of a synchronous generator connected to a power system. We'll assume that the generator is synchronized with the system and the generator is "floating" on the line meaning that there is no input torque to the generator from the prime mover (steam turbine, for example). The power system is supplying a small amount of power to the generator to satisfy the rotational and ohmic losses; thus, the rotor angle δ is close to zero.

To supply electrical power to the power system, an input torque must be supplied from the steam turbine by appropriate steam valving. This is done in increments whereupon each incremental increase in input torque causes the rotor speed to increase slightly thus increasing the rotor angle δ. This incremental loading continues until the generator is supplying the prescribed amount of power to the power system without exceeding the stable operating range of the generator. Example 6B should help to illustrate typical rated, steady-state operation of a synchronous generator.

It is common to express the torque and rotor speed relationship given by (6.53) in terms of rotor angle. From (6.57), we can write

$$\begin{aligned}\frac{d\delta}{dt} &= \frac{d\theta_r}{dt} - \frac{d\theta_{esv}}{dt}\\ &= \omega_r - \omega_e\end{aligned} \tag{6.72}$$

where ω_e is the electrical angular velocity of the power system which is generally assumed to be constant. Thus,

$$\frac{d\omega_r}{dt} = \frac{d^2\delta}{dt^2} \tag{6.73}$$

Therefore, (6.53) can be written

$$T_e - T_L = J\left(\frac{2}{P}\right)\frac{d^2\delta}{dt^2} \tag{6.74}$$

The friction and windage term containing B_m is often neglected. We see that when the torque output, T_e, is greater than the load torque, T_L, the machine momentarily accelerates. When $T_L > T_e$ it momentarily slows.

We have mentioned that the power system engineers assume positive current out of the synchronous generator. With this reversing of the sense of positive current, power and torque are positive for generator action. Thus, the negative sign is deleted from (6.70). Also, the first term on the right-hand side of (6.67) would be negative.

There is one last thing before considering Example 6B; location of the stator poles. For this purpose, we need (2.45) which is repeated here as

$$\text{mmf}_s = \frac{N_s}{2}\sqrt{2}I_s\cos[(\omega_e - \omega)t + \theta_{esi}(0) - \theta(0) - \phi] \tag{6.75}$$

where ω_e is the electrical angular velocity of the stator variables, ω is the electrical angular velocity of the q axis, $\theta(0)$ is the time zero position of the q axis, and ϕ is the position from the q axis to the south pole of the stator. Now, two things;

first (6.75) is for a two-phase stator, but we know that the three phase differs only by the $\frac{3}{2}$ factor, so we are not going to worry about that. Secondly, in the case of the permanent-magnet ac machine, $\theta(0)$ was zero; here $\theta(0)$ is δ. If now we let $\omega = \omega_e$ and $\theta(0) = \delta$, (6.75) becomes

$$\text{mmf}_s^e = \frac{N_s}{2}\sqrt{2}I_s\cos(\theta_{esi}(0) - \delta - \phi_e) \tag{6.76}$$

We are now moving at ω_e, the location of the south pole due to the stator currents is located at

$$\phi_e = \theta_{esi}(0) - \delta \tag{6.77}$$

relative to the q axis or in phase with $\tilde{I}_{as}$, which is rotating at ω_e.

Example 6B ***Phasor diagram of steam turbine generator operation***
Consider a steam turbine generator with the following ratings and parameters:
Rating: 835 MVA (mega volt ampere), Line-to-line voltage: 26 kV
Power factor: 0.85, Poles: 2, Speed: 3600 r/min
Inertia: $J = 0.0658 \times 10^6$ J s^2 (turbine and generator)
Resistances and reactances in ohms:

$$r_s = 2.43 \times 10^{-3} \qquad X_{ls} = 0.1538 \qquad r'_{fd} = 7.5 \times 10^{-4}$$

$$\frac{3}{2}X_{ms} = 1.3032 \qquad X_{ss} = 1.457 \qquad X'_{lfd} = 0.1145$$

For balanced, steady-state, rated operation, calculate $\tilde{E}_a$, E'_x, T_e, and draw the closed phasor diagram and locate the stator and rotor poles.

From (6.67) and since the last term is $\tilde{E}_a$

$$\tilde{E}_a = \tilde{V}_{as} - (r_s + j\omega_e L_{ss})\tilde{I}_{as} \tag{6B.1}$$

The magnitude of the phase current is the MVA per phase divided by the root-mean-square (rms) value of the phase voltage. According to (6.10), the line-to-line voltage divided by $\sqrt{3}$ yields the phase voltage. Therefore,

$$\begin{aligned}|\tilde{I}_{as}| &= \frac{\text{MVA}}{3}\left(\frac{1}{|\tilde{V}_{as}|}\right)\\ &= \frac{835}{3}\left(\frac{\sqrt{3}}{26 \times 10^3}\right) = 18.54\ \text{kA}\end{aligned} \tag{6B.2}$$

The power factor is given as 0.85. We must be careful here. If we were dealing with motor operation and positive current into the machine, then $\tilde{I}_{as}$ would be positioned at $\theta_{esi}(0) = \pm 31.8°$ with the time-zero position of $\tilde{V}_{as}$ at zero degrees. The reactive power is said to be absorbed if $\sin[\theta_{esv}(0) - \theta_{esi}(0)]$ is positive with positive current in the direction of a voltage drop. Since a generator will generally deliver reactive power and since we are assuming positive current into the machine, $\theta_{esi}(0) = 180° - 31.8°$ or 148.2°. Whereupon, power is negative (generator action) and reactive power is also negative (producing reactive power).

In calculating $\widetilde{E}_a$ using (6B.1), we are neglecting r_s since it is small relative to the inductive reactances. Thus, from (6B.1)

$$\widetilde{E}_a = \frac{26 \times 10^3}{\sqrt{3}}\underline{/0°} - (j1.457)18.54 \times 10^3\underline{/180° - 31.8°}$$

$$= 15.03 \times 10^3 + 26.9 \times 10^3\underline{/90 - 31.8} = 37.1 \times 10^3\underline{/38.1°} \tag{6B.3}$$

We can calculate E_x' from (6.71). The torque may be calculated from (6.70)

$$T_e = -\left(\frac{P}{2}\right)\left(\frac{3}{2}\right)\left(\frac{1}{\omega_e}\right)\left(\frac{1}{X_{ss}}\right)E_x'\sqrt{2}V_s \sin\delta$$

$$= -\left(\frac{2}{2}\right)\left(\frac{3}{2}\right)\left(\frac{1}{377}\right)\left(\frac{1}{1.457}\right)(\sqrt{2}\,37.1 \times 10^3)(\sqrt{2}\,15.03 \times 10^3)\sin 38.1°$$

$$= -1.88 \times 10^6 \text{ N m} \tag{6B.4}$$

Power is being supplied to the electric system by the generator. The phasor diagram is shown in Figure 6B.1

From Figure 6B.1, we see that the rotor poles are "pushing" the stator poles in the counterclockwise direction; generator action. It should be pointed out that we can only superimpose the rotor q-, d-variables on the phasor diagram when $\omega_r = \omega_e$.

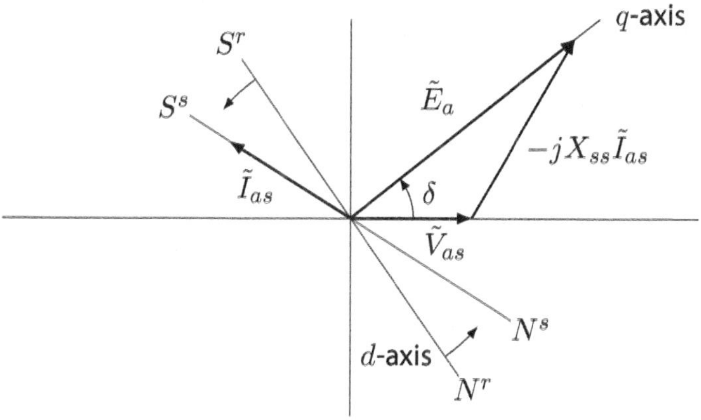

Figure 6B.1 Phasor diagram with r_s neglected for generator action and with the direction of positive current selected to be into the machine.

SP6.3.1 Consider the steam turbine generator given in Example 6B. Determine $\theta_{esi}(0)$ for operation at rated power output with rated power factor and absorbing reactive power. [$\theta_{esi}(0) = 211.8°$]

SP6.3.2 Calculate $\widetilde{E}_a$ for SP6.3.1 and sketch the phasor diagram. [$\widetilde{E}_a = 22.8 \times 10^3 \underline{/87.8^\circ}$ V]

SP6.3.3 Calculate T_e for SP6.3.1 using (6.70). [$T_e = -1.87 \times 10^6$ N m]

SP6.3.4 Why is this mode of operation undesirable? [$\delta = 87.8^\circ$, near the steady-state power limit]

SP6.3.5 Suggest how this undesirable mode of operation can be corrected. [Increase $\widetilde{E}_a$ by increasing I_{fd}]

6.4 Reactive Power and Power-Factor Correction

In Chapter 1, we introduced reactive power, which really isn't power but an interchange of energy to and from the capacitances and inductances of the system. That is, energy is stored for one-half of the cycle and returned during the other one-half. A measure of this energy interchange is calculated using (1.31); that is

$$Q = |\widetilde{V}|\,|\widetilde{I}|\,\sin[\theta_{ev}(0) - \theta_{ei}(0)] \tag{6.78}$$

where positive current is assumed in the direction of voltage drop per phase. Reactive power has the units of volt–ampere reactive (var) and it is said to be "absorbed" if Q is positive and "generated" if Q is negative. In other words, Q is absorbed by an inductance and supplied by a capacitance.

The vast majority of electric power system loads are inductive and the reactive power absorbed by the inductive loads must be generated or supplied by capacitors and/or synchronous generators. How is this "supplying" done? Well, in Example 6B, the steam turbine generator was operating at rated conditions. The phasor diagram depicting this mode of operation is shown in Figure 6B.1. Let us calculate Q for this operation; here

$$Q = 3(15\text{ kV})(18.5\text{ kA})\sin[0 - 148^\circ] = -441\text{ Mvar} \tag{6.79}$$

The generator is supplying both real and reactive power; it is "acting" as a capacitor. How does this happen? Well, the per-phase steady-state voltage equation for a synchronous machine is given by (6.67), which can be written

$$\widetilde{V}_{as} = jX_{ss}\widetilde{I}_{as} + \widetilde{E}_a \tag{6.80}$$

where the stator resistance, r_s, is neglected, and

$$X_{ss} = \omega_e L_{ss} \tag{6.81}$$

also

$$\widetilde{E}_a = \frac{1}{\sqrt{2}}\omega_e\left(\frac{3}{2}\right)L_{ms}I_{fd}^{r}e^{j\delta} \tag{6.82}$$

For this explanation, let us assume that the machine is connected to the power system. If there is no external load or input torque to the shaft; the machine is said to be "floating on the line." Actually, the small amount of power being supplied to the machine from the power system is the $|\widetilde{I}_s|^2 r_s$ and the small friction and windage losses. We will allow this small violation of a floating machine and assume that the rotor angle to be zero. The phasor diagram for this mode of operation is shown in Figure 6.7. In Figure 6.7a, $\widetilde{E}_a$ (I_{fd}^r) is zero; $\widetilde{I}_{as}$ is lagging $\widetilde{V}_{as}$ by 90° and $jX_{ss}\widetilde{I}_{as}$ is equal to $\widetilde{V}_{as}$. The synchronous machine "looks like" an inductance. In Figure 6.7b, I_{fd}^r has been increased to where $\widetilde{I}_{as}$ is negligibly small, sufficient to satisfy the small rotational and ohmic losses. In this case, $\widetilde{E}_a$ is near equal to $\widetilde{V}_{as}$. In Figure 6.7c, $\widetilde{E}_a$ (I_{fd}^r) is increased and $\widetilde{E}_a > \widetilde{V}_{as}$ and, to satisfy (6.80), $\widetilde{I}_{as}$ must lead $\widetilde{V}_{as}$ by 90° since we are neglecting r_s. Here the machine "looks like" a capacitor, delivering reactive power. The mode of operation of the synchronous generator illustrated in Figure 6.7a is said to be under excited, while the mode of operation shown in Figure 6.7c is over excited. If $|\widetilde{I}_{as}|$ is plotted on the y axis and I_{fd}

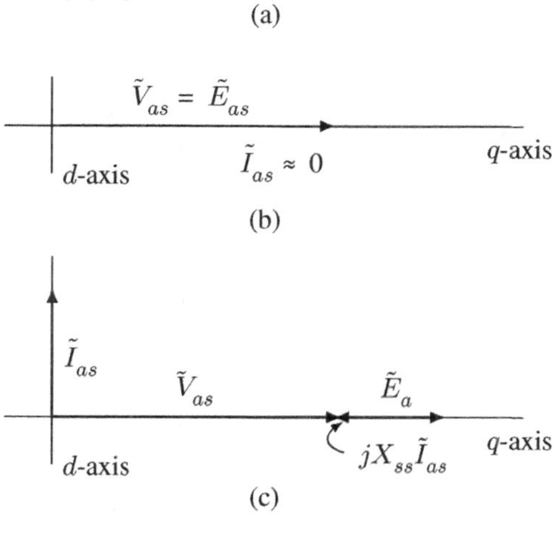

Figure 6.7 (a) I_{fd}^r ($\widetilde{E}_a$) = zero (b) $\widetilde{E}_a = \widetilde{V}_{as}$ (c) $\widetilde{E}_a > \widetilde{V}_{as}$.

or $|\widetilde{E}_a|$ on the x axis, we would obtain the so-called "V" curve. Therefore, the synchronous generators connected to the power system are generally over excited to supply reactive power to the inductive power system loads. A synchronous generator operated without a torque input or load, but instead is operated solely to supply reactive power is referred to as a synchronous condenser.

There is a good reason why synchronous generators are not used to supply all of the reactive power to the inductive loads. We will use Figure 6.8 to help explain this. The elementary transmission system shown in Figure 6.8 could represent single-phase transmission as in rural areas or one phase of a three-phase system. Therein, $\widetilde{V}_G$ and $\widetilde{I}_G$ can be thought of as phase variables of a three-phase generator. The transmission-line resistance and inductance are r_l and x_l, respectively. The load is assumed to be a series RL; R_s and X_s. It is convenient to consider the series load as the parallel equivalent whereupon Figure 6.8 becomes Figure 6.9. In Figure 6.9

$$R_p = \frac{R_s^2 - (jX_s)^2}{R_s} = \frac{R_s^2 + X_s^2}{R_s} \tag{6.83}$$

$$jX_p = -\frac{R_s^2 - (jX_s^2)}{jX_s} = j\frac{R_s^2 + X_s^2}{X_s} \tag{6.84}$$

Figure 6.10 depicts the phasor diagram for the elementary system given in Figure 6.9. With only the generator supplying the reactive power the transmission line must carry the reactive component, $\widetilde{I}_X$, of the load current, $\widetilde{I}_L$. If now we place a capacitor at or near the load the power system becomes that shown in Figure 6.11, and if we select X_C equal to X_p the phasor diagram becomes Figure 6.12. Since the transmission line carries only the real component of the load current, the voltage drop across the transmission line is reduced and thus

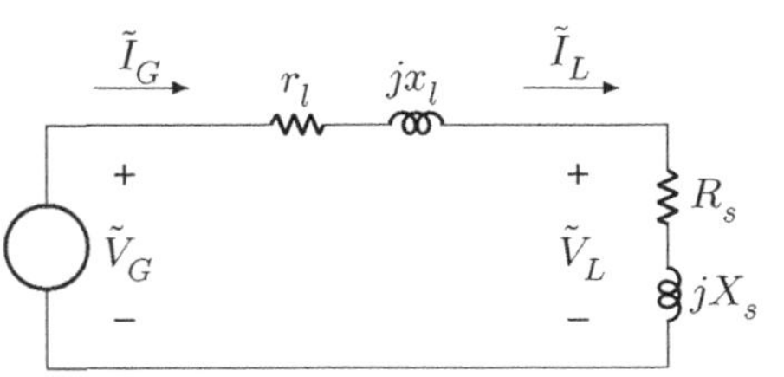

Figure 6.8 One-line diagram of an elementary power system with a series R_s and X_s load.

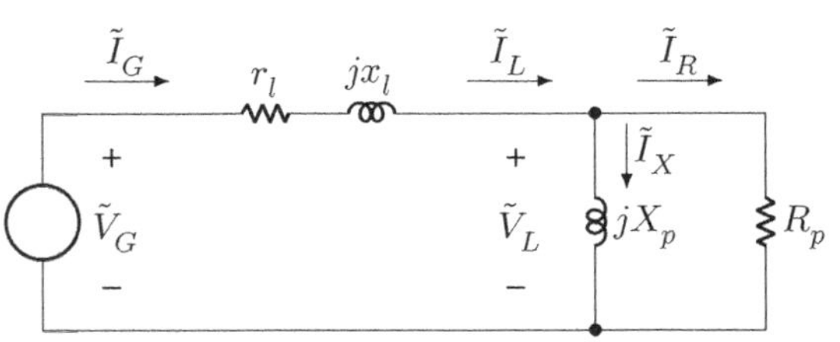

Figure 6.9 One-line diagram of an elementary power system with equivalent parallel connected load.

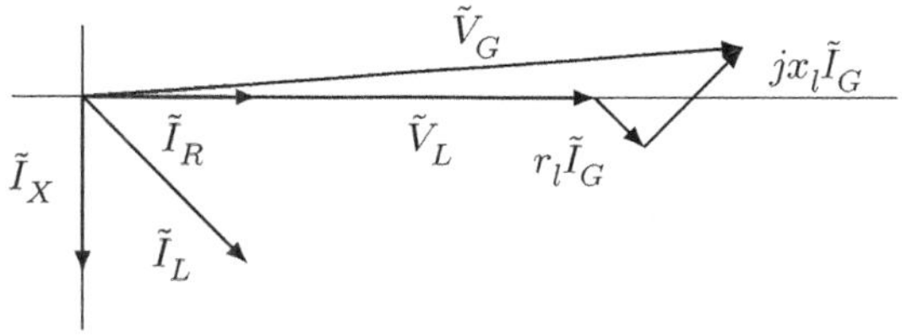

Figure 6.10 Phasor diagram for Figure 6.9.

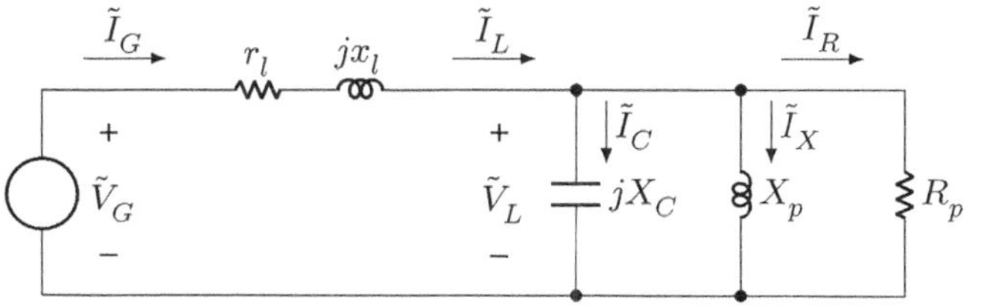

Figure 6.11 One-line diagram of an elementary power system with power-factor correction capacitor in parallel with the load.

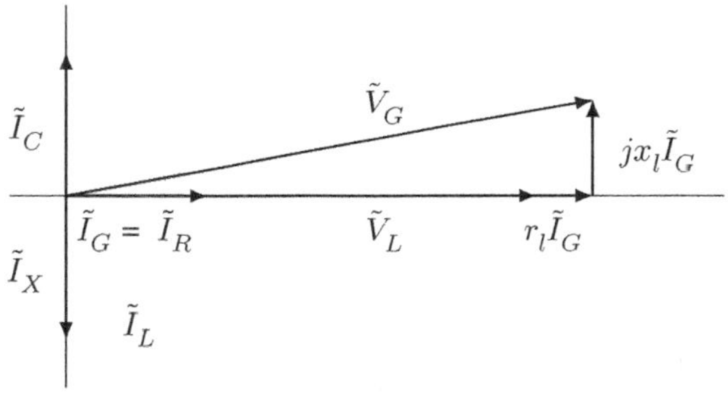

Figure 6.12 Phasor diagram for Figure 6.11 with $X_C = X_p$.

$\tilde{V}_L$ is increased. Some three-phase capacitor banks are connected permanently near the load center; others are automatically "switched in" during heavy load conditions when the load voltage has dropped below a preset level. This common means of load voltage support is referred to as power-factor correction.

Example 6C ***Power-factor correction***

The 60-Hz current supplied to a single-phase load is $\tilde{I}_{L1} = 10\underline{/-30°}$ A and the voltage is $110\underline{/0°}$ V. A second load is connected in parallel and the current supplied to this load is $I_{L2} = 20\underline{/-60°}$ A. (a) Determine the parallel equivalent R_{p1}, X_{p1}, R_{p2}, and X_{p2}. (b) Determine the capacitance that, when connected in parallel with the loads, will return the power factor angle to 30°. (c) Calculate the percent decrease in total load current from (a) to (b).

For (a) let us first write the load currents as

$$\tilde{I}_{L1} = 8.66 - j5\ \text{A} \tag{6C.1}$$

$$\tilde{I}_{L2} = 10 - j17.3\ \text{A} \tag{6C.2}$$

Now,

$$R_{p1} = \frac{110}{8.66} = 12.7\ \Omega \tag{6C.3}$$

$$X_{p1} = \frac{110}{5} = 22\ \Omega\quad \text{(inductive)} \tag{6C.4}$$

$$R_{p2} = \frac{110}{10} = 11\ \Omega \tag{6C.5}$$

$$X_{p2} = \frac{110}{17.3} = 6.3\ \Omega\quad \text{(inductive)} \tag{6C.6}$$

We could have used (6.83) and (6.84) to calculate R_p and X_p. For example,

$$\begin{aligned} Z_{L1} &= \frac{\tilde{V}}{\tilde{I}_{L1}} \\ &= \frac{110\angle 0^\circ}{10\angle -30^\circ} = 11\angle 30^\circ = 9.5 + j5.5\ \Omega \end{aligned} \tag{6C.7}$$

Substituting into (6.83)

$$\begin{aligned} R_{p1} &= \frac{R_s^2 + X_s^2}{R_s} \\ &= \frac{(9.5)^2 + (5.5)^2}{9.5} = 12.7\ \Omega \end{aligned} \tag{6C.8}$$

From (6C.1) and (6C.2), the total load current is

$$\begin{aligned} \tilde{I}_{LTa} &= \tilde{I}_{L1} + \tilde{I}_{L2} \\ &= (8.66 + 10) - j(5 + 17.3) \\ &= 18.66 - j22.3\ \text{A} \end{aligned} \tag{6C.9}$$

For (b) we want the total load current, I_{LTb}, to have a phase angle of -30°, thus

$$|\tilde{I}_{LTb}| = \frac{18.66}{\cos 30^\circ} = 21.55\ \text{A} \tag{6C.10}$$

and

$$\begin{aligned} \tilde{I}_{LTb} &= 18.66 - j21.55 \sin 30^\circ \\ &= 18.66 - j10.8\ \text{A} \end{aligned} \tag{6C.11}$$

Therefore, the parallel capacitor must "supply"

$$|\tilde{I}_{\text{cap}}| = 22.3 - 10.8 = 11.5\ \text{A} \tag{6C.12}$$

whereupon

$$|Z_{\text{cap}}| = \frac{|\tilde{V}|}{|\tilde{I}_{\text{cap}}|} = \frac{110}{11.5} = 9.57\ \Omega \tag{6C.13}$$

Therefore, the capacitance is

$$C = \frac{1}{|Z_{\text{cap}}\omega_e|} = \frac{1}{(9.57)(377)} = 277\ \mu F \tag{6C.14}$$

For (c), the magnitude of the total load current for (a) from (6C.9) is

$$|\tilde{I}_{LTa}| = (18.66^2 + 22.3^2)^{\frac{1}{2}} = 29.1\ \text{A} \tag{6C.15}$$

From (6C.11)

$$|\tilde{I}_{LTb}| = (18.66^2 + 10.8^2)^{\frac{1}{2}} = 21.6\ \text{A} \tag{6C.16}$$

The percent decrease in the magnitude of the load (line) current is

$$\%\text{decrease} = \left(\frac{29.1 - 21.6}{29.1}\right) 100 = 25.8\% \tag{6C.17}$$

Although this example is perhaps academic, a similar situation occurred when in the mid-1950s the eldest author lived with his young family in veteran's housing at a distance from the power source. Window air conditioners were being installed by many residents that were cheaply made with a small power factor. The large reactive load current caused the voltage at the housing area (World War II converted hospital barracks) to drop to where the motor in the air conditioners would stop, unable to handle the load of the fan and compressor. It was a hot Nebraska summer. Finally, banks of capacitors were installed by the utility. Later, restrictions were placed on air conditioner manufacturers to improve the power factor.

SP6.4.1 Show that (6.78) is $|\tilde{I}|^2 X$ for a series resistive and reactive circuit.

SP6.4.2 Derive (6.83) and (6.84).

SP6.4.3 Repeat Example 6C(b) if $I_{LTb} = 18.66\underline{/0^\circ}$ A. [$C = 538\ \mu$F]

SP6.4.4 Locate the stator and rotor poles in Figure 6.7. [N^r at $-\frac{\pi}{2}$ in all; (a) S^s at $-\frac{\pi}{2}$; (b) I_{as} assumed zero; (c) S^s at $\frac{\pi}{2}$]

6.5 Per Unit System

The per unit system is a means of normalizing the system parameters and variables to a specified base power for the entire system being considered. The base voltage is generally selected as the rated voltage of each system component; generator, transmission line, load. The power system engineers are especially fond of the per unit system since it is far more convenient to express the voltage of the synchronous machine in Example 6B as one per unit rather than 26 000 volts or 26 kV. More importantly, it allows a means of dealing with different voltage levels with all transformers considered ideal. Base power for a power system would generally be much larger than that of a single machine, and it is generally given in volt–amperes rather than watts. If we were to per unitize just the synchronous machine in Example 6B, the power base would be 835 MVA and the line-to-line voltage base would be 26 kV. The base phase voltage would be $\frac{1}{\sqrt{3}}$(26 kV). Since the 26 kV is an rms value, base peak voltage would be $\sqrt{2}$ times the rms values. Once the base power is specified, the base current will depend upon the voltage base selected for each system component. This can be confusing if the base voltage is not defined as either a phase or line-to-line voltage; and either rms or peak value.

The parameters of the power system and of the machine are commonly per-phase quantities. With the phase rms voltage and current as base quantities, base power may be expressed

$$P_{\text{base}} = 3V_{\text{base}}I_{\text{base}} \tag{6.85}$$

where P_{base} is base system power and V_{base} is the rated phase voltage of the system component being considered. The impedance base is then

$$Z_{\text{base}} = \frac{V_{\text{base}}}{I_{\text{base}}} = \frac{3V_{\text{base}}^2}{P_{\text{base}}} \tag{6.86}$$

where I_{base} is determined from (6.85) once V_{base} is established. Although P_{base} is the same throughout the system being considered, V_{base} and I_{base} will change for each system voltage level.

Base torque is

$$T_{\text{base}} = \frac{P_{\text{base}}}{\left(\frac{2}{P}\right)\omega_{\text{base}}} \tag{6.87}$$

where ω_{base} is the electrical angular velocity corresponding to rated frequency. Per unitizing the q- and d-voltage equations of a generator is rather involved. We will leave it to advance study [1].

The steady-state power output P_e may be expressed in per unit by starting with (6.70) and realizing that for positive current out of the machine the negative sign on the right-hand side of (6.70) becomes positive and stable generator operation occurs $0 < \delta < \frac{\pi}{2}$. Also we must recall that steady-state power is torque times the actual rotor speed ω_{rm}. The per unit electrical power output, positive for generator action, becomes

$$P_e = \frac{|\widetilde{E}_a| V_s}{X_{ss}} \sin \delta \tag{6.88}$$

The torque-rotor angle relationship is given by (6.74). We will neglect the term containing B_m since this term is very small in comparison to the power level of the power system. Thus, for generator action (6.74) becomes

$$T_e = -J \left(\frac{2}{P} \right) p^2 \delta + T_I \tag{6.89}$$

where $p^2 = \frac{d^2}{dt^2}$ and T_e is positive for generator action and T_I is positive for an input torque to the generator shaft from the prime mover (steam turbine). Now, to per unitize (6.89), we can write

$$\frac{T_e}{T_{\text{base}}} = -\frac{J\left(\frac{2}{P}\right) p^2 \delta}{T_{\text{base}}} + \frac{T_I}{T_{\text{base}}} \tag{6.90}$$

The power system engineer prefers to express (6.90) in per unit as

$$T_e = -\frac{2H}{\omega_{\text{base}}} p^2 \delta + T_I \tag{6.91}$$

where, in seconds

$$H = \left(\frac{1}{2} \right) \left(\frac{2}{P} \right) \frac{J \omega_{\text{base}}}{T_{\text{base}}} \tag{6.92}$$

Example 6D ***Consider the elementary power system shown in Figure 6D.1*** (a) Use $P_{\text{base}} = 1000$ MVA and determine V_{base}, I_{base}, and Z_{base} for each of the voltage levels. (b) Per unitize the parameters of the steam turbine generator given in Example 6B. Use $P_{\text{base}} = 835$ MVA with the rms phase voltage as the base voltage.

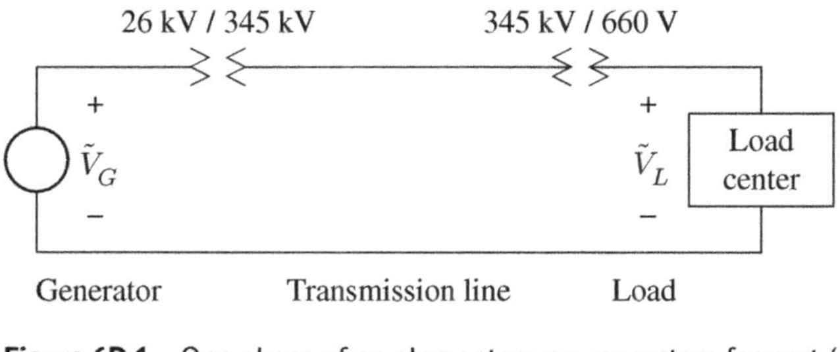

Figure 6D.1 One phase of an elementary power system for part (a). Voltage levels are given in line-to-line values.

For part (a), since the ratings are given in line-to-line, the phase voltage ratings (V_{base}) are obtained by dividing by $\sqrt{3}$. The power base per phase is $\frac{1000}{3}$MVA or 333 MVA. For the generator given in Example 6B

$$V_{\text{base}} = \frac{26\text{ kV}}{\sqrt{3}} = 15\text{ kV} \tag{6D.1}$$

$$I_{\text{base}} = \frac{333 \times 10^6}{(15 \times 10^3)} = 22\text{ kA} \tag{6D.2}$$

$$Z_{\text{base}} = \frac{V_{\text{base}}}{I_{\text{base}}} = \frac{15\text{ kV}}{22\text{ kA}} = 0.682\ \Omega \tag{6D.3}$$

Note that we are using P_{base} different than the machine rating. For the transmission line

$$V_{\text{base}} = \frac{345\text{ kV}}{\sqrt{3}} = 199\text{ kV} \tag{6D.4}$$

$$I_{\text{base}} = \frac{333 \times 10^6}{199 \times 10^3} = 1.673\text{ kA} \tag{6D.5}$$

$$Z_{\text{base}} = \frac{199\text{ kV}}{1.673\text{ kA}} = 118.9\ \Omega \tag{6D.6}$$

For the load

$$V_{\text{base}} = \frac{660}{\sqrt{3}} = 382\text{ V} \tag{6D.7}$$

$$I_{\text{base}} = \frac{333 \times 10^6}{382} = 872\text{ kA} \tag{6D.8}$$

$$Z_{\text{base}} = \frac{382}{872 \times 10^3} = 0.438 \times 10^{-3}\ \Omega \tag{6D.9}$$

For (b)

$$V_{\text{base}} = \frac{1}{\sqrt{3}} 26\text{ kV} = 15.03\text{ kV} \tag{6D.10}$$

$$\begin{aligned} I_{\text{base}} &= \frac{P_{\text{base}}}{3V_{\text{base}}} \\ &= \frac{835\text{ MVA}}{(3)15.03\text{ kV}} = 18.5\text{ kA} \end{aligned} \tag{6D.11}$$

$$\begin{aligned} Z_{\text{base}} &= \frac{V_{\text{base}}}{I_{\text{base}}} \\ &= \frac{15.03\ \text{ kV}}{18.3\ \text{ kA}} = 0.812\ \Omega \end{aligned} \tag{6D.12}$$

$$T_{\text{base}} = \frac{P_{\text{base}}}{\left(\frac{P}{2}\right)\omega_{\text{base}}}$$
$$= \frac{835 \text{ MVA}}{\left(\frac{2}{2}\right)(377)} = 2.215 \times 10^6 \text{ N m} \tag{6D.13}$$

The per unit parameters are obtained by dividing the parameters given in ohms in Example 6B by Z_{base}. The following listing gives these parameters in ohms and in per unit (pu)

$$r_s = 2.43 \times 10^{-3}\ \Omega,\ \ 3 \times 10^{-3} \text{ pu} \qquad X_{ls} = 0.1538\ \Omega,\ \ 0.189 \text{ pu}$$
$$\tfrac{3}{2}X_{ms} = 1.3032\ \Omega,\ \ 1.6 \text{ pu} \qquad X_{ss} = 1.457\ \Omega,\ \ 1.794 \text{ pu}$$
$$r'_{fd} = 7.5 \times 10^{-4}\ \Omega,\ \ 9.236 \times 10^{-4} \text{ pu} \qquad X'_{lfd} = 0.1145\ \Omega,\ \ 0.141 \text{ pu}$$

In part (a) we used $P_{\text{base}} = 1000$ MVA, the above parameters are per unitized on a P_{base} of 835 MVA. To convert the above per unit values to a 1000 MVA base, we would multiply by 1000/835 if V_{base} remains the same.

Now from (6.92) with $J = 0.0658 \times 10^6$ J s^2

$$H = \left(\frac{1}{2}\right)\left(\frac{2}{2}\right)\frac{J\omega_{\text{base}}}{T_{\text{base}}}$$
$$= \left(\frac{1}{2}\right)\frac{0.0658 \times 10^6(377)}{2.215 \times 10^6} = 5.6 \text{ seconds} \tag{6D.14}$$

SP6.5.1 Some prefer to use per unit inductance rather than per unit reactance. Determine base inductance for Example 6D part (b). [$L_{\text{base}} = 2.15$ mH]

SP6.5.2 If the rms phase voltage is used to per unitize q_s^r- and d_s^r-voltages, determine the per unit value of maximum v_{qs}^r or v_{ds}^r. [$\sqrt{2}$ pu]

6.6 Discussion of Transient Stability

Large-excursion stability, which is commonly referred to as transient stability, is of major concern to the power system engineer. Although we will not get too involved, we can use the material in Section 6.3 to describe some of the major aspects of this phenomenon. As we have mentioned, the vast majority of electric power is generated by synchronous generators and, for multiple generators to supply power to an electric system, all of the generators must be synchronized with the electric system; in other words, all synchronous generators connected to an electric system must rotate at the same electrical angular velocity. Large areas of the power grid of the United States are "synchronously tied together." Since there are hundreds of large synchronous generators connected to a large power grid, we can understand the concern of the power system engineer.

6.6.1 Three-phase Fault

Our purpose is to describe large-excursion stability. To do this, we'll consider the system in Figure 6.13. Although this example may be considered academic, it serves our need without becoming overly involved. Initially, the steam turbine generator that we considered in Example 6B is delivering rated MVA at rated power factor with essentially all of the power being delivered to the power grid. A three-phase fault occurs on the line feeding the negligibly small local load near the generator. This is simulated by stepping the three-phase voltages at the bus to zero. These voltages are held at zero until the circuit breaker, CB, opens, clearing the fault from the bus. Whereupon it is assumed that the voltage at the bus steps back to normal magnitude, phase, and frequency due to the robustness or "stiffness" of the large system.

Figures 6.14 and 6.15 illustrate the dynamic performance of the steam turbine generator considered in Example 6B during and following the three-phase fault. There are a host of things that we need to explain regarding these computer traces. In Figure 6.14, i_{as}, T_e, ω_r, and δ the power angle are plotted. From Example 6B the initial δ is $\delta_0 = 38.1\,°$.

The dynamic torque-angle characteristics, with the electrical transients neglected, during and following the fault, are illustrated in Figure 6.15. The generator is initially operating in the steady state with the torque input $T_I = (0.85)2.22\times10^6$ N m and $E_x' = (2.48)\sqrt{\frac{2}{3}}26$ kV. Recall we are now power system engineers and we changed the assumed direction of positive current; thus, the input torque is T_I and it is positive for generator action. With this change, stable steady-state operation occurs on the positive-slope part of the T_e versus δ plot $\left(-\frac{\pi}{2} < \delta < \frac{\pi}{2}\right)$; however, positive T_e occurs from $0 < \delta < \pi$.

In the more sophisticated present-day transient stability programs, the q- and d-circuits are used for the machines and the transmission network with the machine represented in their rotor reference frame and the network represented in the synchronously rotating reference frame. The electric transient are neglected in the stator windings of the machines and in the network, thereby eliminating the

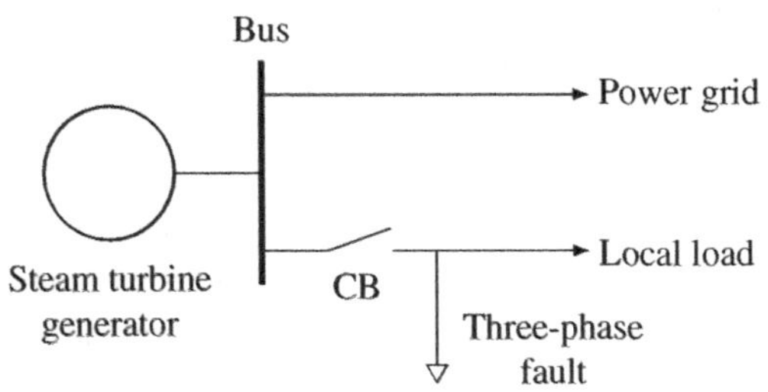

Figure 6.13 One-line diagram of system configuration for three-phase fault.

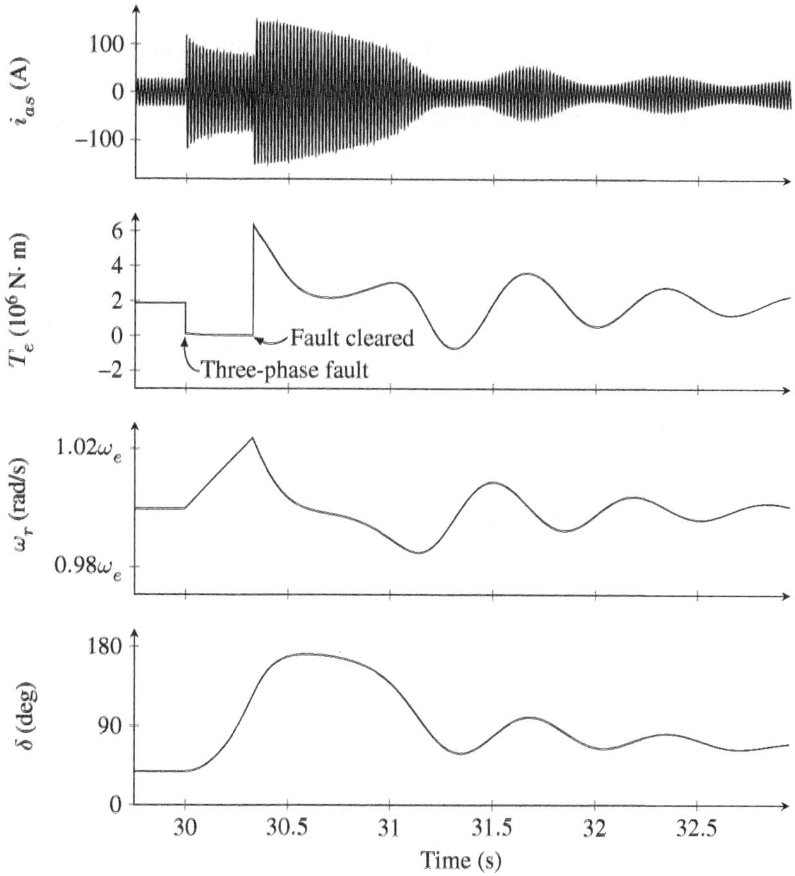

Figure 6.14 Dynamic performance of the steam turbine generator during a three-phase fault at the terminals predicted with stator and transmission line electric transients neglected.

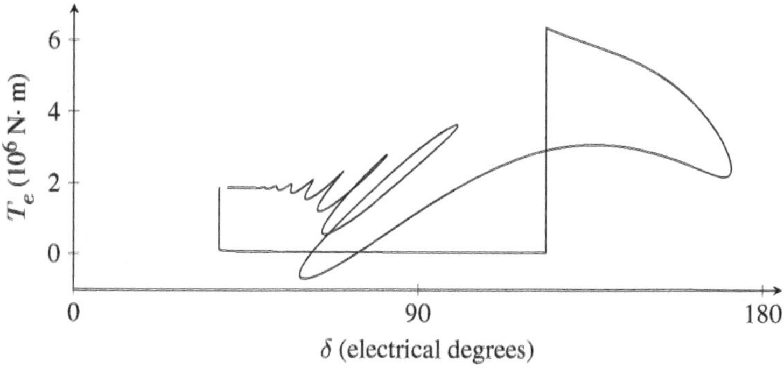

Figure 6.15 Torque versus angle characteristics for the study shown in Figure 6.14.

need to integrate for each inductance in the stator and network circuit equations while maintaining sufficient accuracy for transient stability studies [2]. If the electric transients were taken into account a rapidly decaying dc offset would be present in stator and network currents following an electrical disturbance. The decaying dc offset in the stator currents gives rise to a short-term 60-Hz decaying pulsation in T_e [1].

As we mentioned in Section 6.3, damper windings, which are short-circuited rotor windings, are needed to damp rotor oscillations after a disturbance. Although we included these windings in the simulation of the steam turbine generator for studies shown in Figures 6.14 and 6.15, we will not include these windings in our analysis here since the effects of damper windings on transient stability are minimal.

Now, (6.57) defines δ and as we have mentioned, the power system is large and is often considered to have infinite electrical and mechanical inertia. Although the generator connected to the power system bus can deliver power to the system, the voltage and frequency of a large power system are essentially independent of the generator operation. This large power system bus is often referred to as an "infinite bus." Now, it can be argued that no such thing exists; clearly, a micro grid is not an infinite inertia system; nevertheless, we'll assume an infinite bus in this discussion whereupon θ_{esv} is constant and since we are now power system engineers, (6.74) becomes

$$T_I - T_e = J\left(\frac{2}{P}\right)\frac{d^2\delta}{dt^2} \tag{6.93}$$

When the input torque, T_I, from the steam turbine is greater than the output torque, T_e, or power output, the machine accelerates; when $T_e > T_I$ it slows. Equation (6.93) is often referred to as the "swing equation." It is the dynamic relationship between the "swing" of the rotor of a generator relative to a large power system. That is, each generator connected to the power system would have its own swing equation made up of its own T_e, T_I, inertia, and rotor angle δ.

Let's now get back to Figures 6.14 and 6.15. As mentioned, the three-phase fault is simulated by stepping the terminal voltages to zero; therefore, power cannot be transferred from the generator to the power grid. Thus, the input torque T_I accelerates the rotor as shown in the plot of ω_r in Figure 6.14 where ω_e, in the plot of ω_r, is the electrical angular velocity of the power system. The fault is removed by opening the CBs to the local load in 0.334 seconds whereupon the machine is reconnected to the power system and the oscillations in rotor speed subside due to the action of the damper windings. Had the fault been allowed to remain slightly longer and T_e had become less than T_I, the machine would again accelerate and would not have returned to synchronous operation with the grid without going through a time consuming resynchronization, or by what is called "slipping poles," which is not a common practice since system and/or machine damage can result.

Now, we should mention that a sudden three-phase fault is rare and a "bolted" three-phase fault near the machine is even rarer; however, it is often used in transient stability studies since it is a measure of the robustness of the system to faults. The dynamic torque-angle characteristics due to the three-phase fault shown in Figure 6.15 are shown in the plot of T_e versus δ from Figure 6.14. Now, we understand that the torque or power from the machine is essentially zero during the fault since the voltage is zero. The goal is to clear the fault and reconnect the machine to the grid and maintain synchronization of the machine with the grid. As we mentioned, synchronism is maintained if the fault is isolated by opening the CB feeding the fault in 0.334 seconds. Thus, the so-called "critical-clearing time" is 0.334 seconds and the "critical-clearing angle" which is the rotor displacement at the critical-clearing angle is approximately 128 electrical degrees.

Let's "fast-backwards" to the early twentieth century when power systems were being built, and the electric industry was growing rapidly and yet computers were years away. It appears that even at the start of World War II, there were less than 20 integrators available in the United States and these were mechanical. Nevertheless, with only slide rules, the engineers were faced with the problem of predicting transient stability. Fortunately, Doherty and Nickle derived an approximate transient torque-angle characteristic that was sufficiently accurate when used with the "equal-area criterion" to predict transient stability [3]. We are not going to dwell on a technique that has long since given way to modern-day computation; however, it is interesting that Doherty and Nickle's method is based on the fact that during the "first swing" of the rotor, which determines stable or unstable operation, the flux linkages of circuits that are largely inductive with a relative small resistance will tend to remain constant. Starting with this assumption, they derived a transient torque-angle curve for the first rotor swing that was easy to determine and use. We should also mention in passing that during the 1930s and 1940s, point-to-point integration for transient stability studies was conducted using a "network calculator" which was a physical mimic of the main high-voltage part of the power system. This was an inconvenient time-consuming device that rapidly gave way to the digital computer in the 1960s.

Example 6E ***Calculation of ω_r and δ at the time of fault clearing***

Consider Figures 6.14 and 6.15. Calculate the rotor speed, ω_r and rotor angle, δ, at the clearing time t_c of 0.334 seconds after the occurrence of the three-phase bolted fault when the synchronous generator is reconnected to the system. The input torque is

$$T_I = (0.85)2.22 \times 10^6 = 1.89 \times 10^6 \text{ N m} \tag{6E.1}$$

This input shaft torque remains on the generator for the duration of the fault, 0.334 seconds. For generator action, the power engineers write (6.53) as

$$T_I - T_e = J\left(\frac{2}{P}\right)\frac{d\omega_r}{dt} \tag{6E.2}$$

where T_L has become $-T_I$, T_e has become positive for generation action, and B_m is neglected. Solving for ω_r yields

$$\begin{aligned}\omega_r &= \frac{P}{2J}\int (T_I - T_e)dt \\ &= \frac{P}{2J}\int_0^{t_c} (T_I - T_e)d\xi \end{aligned} \tag{6E.3}$$

where t_c is the clearing time, $J = 0.0658\times 10^6$, and $P = 2$. Now, T_e is essentially zero during the time the fault is on the system; thus, (6E.3) becomes

$$\begin{aligned}\omega_r &= \frac{(2)(1.89\times 10^6)}{(2)(0.0658\times 10^6)}\ \xi\big|_0^{0.344} + 377 \\ &= 9.59 + 377 = 386.6\ \text{rad/s}\end{aligned} \tag{6E.4}$$

This is 1.025 ω_e.

Now, (6.93) may be written

$$\frac{d^2\delta}{dt^2} = \frac{2J}{P}(T_I - T_e) \tag{6E.5}$$

From Example 6B, the initial rotor angle is $\delta_0 = 38.1°$ and during the fault $T_e = 0$; thus

$$\begin{aligned}\delta &= \frac{P}{2J}\int_0^{t_c}\int_0^{t} T_I d\tau dt + \delta_0 \\ &= \left(\frac{P}{2J}\right)(T_I)\int_0^{t_c} \xi d\xi + \delta_0 \\ &= \left[\left(\frac{P}{2J}\right)\frac{T_I}{2}\ \xi^2\Big|_0^{0.334}\right]\frac{360°}{2\pi} + \delta_0 \\ &= \left[\frac{2(1.89\times 10^6)(0.334)^2}{(2)(0.0658\times 10^6)(2)}\right]\frac{360°}{2\pi} + 38.1° = 129.9°\end{aligned} \tag{6E.6}$$

Note that t_c is the clearing time and that the $\frac{360°}{2\pi}$ multiplier is to convert radians to degrees.

SP6.6.1 Why is T_e during the three-phase not exactly zero. [$3|\tilde{I}_{as}|^2 r_s$]

SP6.6.2 Assume $|I_{as}| = 40$ kA during the fault. Calculate T_e. [0.031×10^6 N m]

6.7 Problems

6.1 Consider the steam turbine generator in Example 6B. Assume it is operating as a motor with rated power input and rated leading power factor. Draw the phasor diagram with all the quantities that are shown in Figure 6B.1.

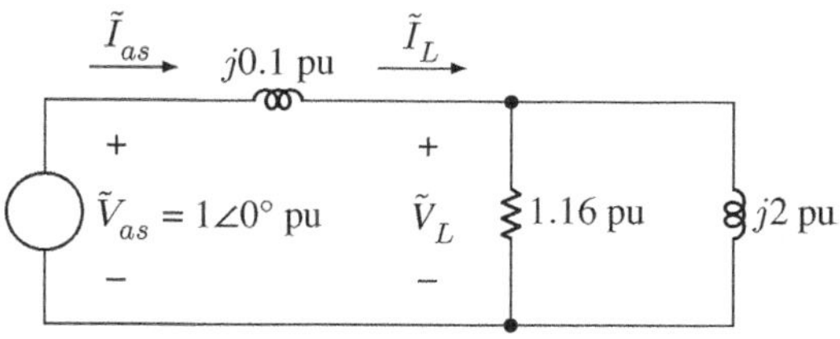

Figure 6.16 Per unit of Figure 6D.1.

6.2 Repeat Problem 6.1 for motor action with rated power factor lagging.

6.3 Consider Example 6C. Rework part (a) in per unit with the power base of 5.5 kVA. Check your work by comparing to the ohmic values given by (6C.3)–(6C.6) and current value given by (6C.9).

6.4 The transmission line parameters in Figure 6D.1 are $r_t = 0.5\ \Omega$ and $x_t = 10\ \Omega$ (inductive). The load current is $755 - j436$ kA at rated phase voltage of $382\underline{/0°}$. Per unitize the elementary power system using the base values determined in Example 6D part (a). Use the system turbine generator considered Example 6D, part (b) as the source.

6.5 The elementary power system shown in Figure 6D.1 is shown in per unit on an 835 MVA three-phase power base in Figure 6.16 below. (a) Calculate $\tilde{V}_L$. (b) Capacitors are distributed throughout the load center to where the load appears only resistive. Recalculate $\tilde{V}_L$.

6.6 Assume that in part (a) of Problem 6.5 that $\tilde{I}_{as} = 1\underline{/-30°}$ and in part (b) $\tilde{I}_{as} = 0.866\underline{/0°}$. Calculate per unit $\tilde{E}_a$ and P_e for both cases.

6.7 Start with (6.70) and obtain the expression for steady-state electrical output power in per unit given by (6.88).

References

1 Krause, P.C. (1986). *Analysis of Electric Machinery*. McGraw-Hill.
2 Krause, P.C., Nozari, F., Skvarenina, T.L., and Olive, D.W. (1979). The theory of neglecting stator transients. *IEEE Trans. Power Apparatus Syst.* 98: 141–148.
3 Doherty, R.E. and Nickle, C.A. (1927). Synchronous machines—III, torque-angle characteristics under transient conditions. *AIEE Trans.* 46: 1–8.

Appendix A

Abbreviations, Constants, Conversions, and Identities

Term	Abbreviation
Alternating current	ac
Ampere	A
Ampere-turn	At
Coulomb	C
Direct current	dc
Electromotive force	emf
Foot	ft
Gauss	G
Gram	g
Henry	H
Hertz	Hz
Horsepower	hp
Inch	in
Joule	J
Kilogram	kg
Kilovar	kvar
Kilovolt	kV
Kilovoltampere	kVA
Kilowatt	kW
Magnetomotive force	mmf
Maxwell	Mx
Megawatt	MW
Meter	m

Introduction to the Analysis of Electromechanical Systems, First Edition.
Paul C. Krause, Oleg Wasynczuk, and Timothy O'Connell.

Term	Abbreviation
Microfarad	μF
Millihenry	mH
Newton	N
Newton meter	N · m
Oersted	Oe
Pound	lb
Poundal	pdl
Power factor	pf
Pulse-width modulation	PWM
Radian	rad
Revolution per minute	r/min (rpm)
Root mean square	rms
Second	s
Voltampere reactive	var
Volt	V
Voltampere	VA
Watt	W
Weber	Wb

A. Constants and Conversion Factors

Permeability of free space	$\mu_0 = 4\pi \times 10^{-7}$ Wb/A m
Permittivity of free space	$\epsilon_0 = 8.854 \times 10^{-12}$ C^2/N m^2
Acceleration of gravity	$g = 9.807$ m/s^2
Length	1 m = 3.218 ft = 39.37 in
Mass	1 kg = 0.0685 slug = 2.205 lb (mass)
Force	1 N = 0.225 lb = 3.6 oz
Torque	1 N m = 0.738 lb ft
Energy	1 J (W s) = 0.738 lb ft
Power	1 W = 1.341×10^{-3} hp
Moment of inertia	1 kg m^2 = 0.738 slug ft^2 = 23.7 lb ft^2
Magnetic flux	1 Wb = 10^8 Mx (lines)
Magnetic flux density	1 Wb/m^2 = 10 000 G = 64.5 klines/in^2
Magnetizing force	1 At/m = 0.0254 At/in = 0.0126 Oe

B. Trigonometric Identities

(I-1) $e^{j\alpha} = \cos\alpha + j\sin\alpha$

(I-2) $a\cos x + b\sin x = \sqrt{a^2+b^2}\cos(x+\phi)\ \phi = \tan^{-1}(-b/a)$

(I-3) $\cos^2 x + \sin^2 x = 1$

(I-4) $\sin 2x = 2\sin x\cos x$

(I-5) $\cos 2x = \cos^2 x - \sin^2 x = 2\cos^2 x - 1 = 1 - 2\sin^2 x$

(I-6) $\cos x\cos y = \frac{1}{2}\cos(x+y) + \frac{1}{2}\cos(x-y)$

(I-7) $\sin x\sin y = \frac{1}{2}\cos(x-y) - \frac{1}{2}\cos(x+y)$

(I-8) $\sin x\cos y = \frac{1}{2}\sin(x+y) + \frac{1}{2}\sin(x-y)$

(I-9) $\cos(x\pm y) = \cos x\cos y \mp \sin x\sin y$

(I-10) $\sin(x\pm y) = \sin x\cos y \pm \cos x\sin y$

(I-11) $\cos^2 x + \cos^2\left(x - \frac{2}{3}\pi\right) + \cos^2\left(x + \frac{2}{3}\pi\right) = \frac{3}{2}$

(I-12) $\sin^2 x + \sin^2\left(x - \frac{2}{3}\pi\right) + \sin^2\left(x + \frac{2}{3}\pi\right) = \frac{3}{2}$

(I-13) $\sin x\cos x + \sin\left(x - \frac{2}{3}\pi\right)\cos\left(x - \frac{2}{3}\pi\right) + \sin\left(x + \frac{2}{3}\pi\right)\cos\left(x + \frac{2}{3}\right) = 0$

(I-14) $\cos x + \cos\left(x - \frac{2}{3}\pi\right) + \cos\left(x + \frac{2}{3}\pi\right) = 0$

(I-15) $\sin x + \sin\left(x - \frac{2}{3}\pi\right) + \sin\left(x + \frac{2}{3}\pi\right) = 0$

(I-16) $\sin x\cos y + \sin\left(x - \frac{2}{3}\pi\right)\cos\left(y - \frac{2}{3}\pi\right) + \sin\left(x + \frac{2}{3}\pi\right)\cos\left(y + \frac{2}{3}\pi\right)$
$= \frac{3}{2}\sin(x-y)$

(I-17) $\sin x\sin y + \sin\left(x - \frac{2}{3}\pi\right)\sin\left(y - \frac{2}{3}\pi\right) + \sin\left(x + \frac{2}{3}\pi\right)\sin\left(y + \frac{2}{3}\pi\right)$
$= \frac{3}{2}\cos(x-y)$

(I-18) $\cos x\sin y + \cos\left(x - \frac{2}{3}\pi\right)\sin\left(y - \frac{2}{3}\pi\right) + \cos\left(x + \frac{2}{3}\pi\right)\sin\left(y + \frac{2}{3}\pi\right)$
$= -\frac{3}{2}\sin(x-y)$

(I-19) $\cos x\cos y + \cos\left(x - \frac{2}{3}\pi\right)\cos\left(y - \frac{2}{3}\pi\right) + \cos\left(x + \frac{2}{3}\pi\right)\cos\left(y + \frac{2}{3}\pi\right)$
$= \frac{3}{2}\cos(x-y)$

(I-20) $\sin x\cos y + \sin\left(x + \frac{2}{3}\pi\right)\cos\left(y - \frac{2}{3}\pi\right) + \sin\left(x - \frac{2}{3}\pi\right)\cos\left(y + \frac{2}{3}\pi\right)$
$= \frac{3}{2}\sin(x+y)$

(I-21) $\sin x \sin y + \sin\left(x + \frac{2}{3}\pi\right)\sin\left(y - \frac{2}{3}\pi\right) + \sin\left(x - \frac{2}{3}\pi\right)\sin\left(y + \frac{2}{3}\pi\right)$
$= -\frac{3}{2}\cos(x + y)$

(I-22) $\cos x \sin y + \cos\left(x + \frac{2}{3}\pi\right)\sin\left(y - \frac{2}{3}\pi\right) + \cos\left(x - \frac{2}{3}\pi\right)\sin\left(y + \frac{2}{3}\pi\right)$
$= \frac{3}{2}\sin(x + y)$

(I-23) $\cos x \cos y + \cos\left(x + \frac{2}{3}\pi\right)\cos\left(y - \frac{2}{3}\pi\right) + \cos\left(x - \frac{2}{3}\pi\right)\cos\left(y + \frac{2}{3}\pi\right)$
$= \frac{3}{2}\cos(x + y)$

Index

Introduction to the Analysis of Electromechanical Systems, First Edition.
Paul C. Krause, Oleg Wasynczuk, and Timothy O'Connell.

d

e

f

h

i

k

l

m

n

o

p

r

IEEE Press Series on Power and Energy Systems
Ganesh Kumar Venayagamoorthy, Series Editor

Series Editor: Ganesh Kumar Venayagamoorthy, Clemson University, Clemson, South Carolina, USA.

The mission of the IEEE Press Series on Power and Energy Systems is to publish leading-edge books that cover a broad spectrum of current and forward-looking technologies in the fast-moving area of power and energy systems including smart grid, renewable energy systems, electric vehicles and related areas. Our target audience includes power and energy systems professionals from academia, industry and government who are interested in enhancing their knowledge and perspectives in their areas of interest.

1. *Electric Power Systems: Design and Analysis, Revised Printing*
   Mohamed E. El-Hawary
2. *Power System Stability*
   Edward W. Kimbark
3. *Analysis of Faulted Power Systems*
   Paul M. Anderson
4. *Inspection of Large Synchronous Machines: Checklists, Failure Identification, and Troubleshooting*
   Isidor Kerszenbaum
5. *Electric Power Applications of Fuzzy Systems*
   Mohamed E. El-Hawary
6. *Power System Protection*
   Paul M. Anderson
7. *Subsynchronous Resonance in Power Systems*
   Paul M. Anderson, B.L. Agrawal, J.E. Van Ness
8. *Understanding Power Quality Problems: Voltage Sags and Interruptions*
   Math H. Bollen
9. *Analysis of Electric Machinery*
   Paul C. Krause, Oleg Wasynczuk, and S.D. Sudhoff
10. *Power System Control and Stability, Revised Printing*
    Paul M. Anderson, A.A. Fouad
11. *Principles of Electric Machines with Power Electronic Applications*, Second Edition
    Mohamed E. El-Hawary
12. *Pulse Width Modulation for Power Converters: Principles and Practice*
    D. Grahame Holmes and Thomas Lipo
13. *Analysis of Electric Machinery and Drive Systems*, Second Edition
    Paul C. Krause, Oleg Wasynczuk, and S.D. Sudhoff
14. *Risk Assessment for Power Systems: Models, Methods, and Applications*
    Wenyuan Li

15. *Optimization Principles: Practical Applications to the Operations of Markets of the Electric Power Industry*
Narayan S. Rau

16. *Electric Economics: Regulation and Deregulation*
Geoffrey Rothwell and Tomas Gomez

17. *Electric Power Systems: Analysis and Control*
Fabio Saccomanno

18. *Electrical Insulation for Rotating Machines: Design, Evaluation, Aging, Testing, and Repair*
Greg C. Stone, Edward A. Boulter, Ian Culbert, and Hussein Dhirani

19. *Signal Processing of Power Quality Disturbances*
Math H. J. Bollen and Irene Y. H. Gu

20. *Instantaneous Power Theory and Applications to Power Conditioning*
Hirofumi Akagi, Edson H. Watanabe, and Mauricio Aredes

21. *Maintaining Mission Critical Systems in a 24/7 Environment*
Peter M. Curtis

22. *Elements of Tidal-Electric Engineering*
Robert H. Clark

23. *Handbook of Large Turbo-Generator Operation and Maintenance*, Second Edition
Geoff Klempner and Isidor Kerszenbaum

24. *Introduction to Electrical Power Systems*
Mohamed E. El-Hawary

25. *Modeling and Control of Fuel Cells: Distributed Generation Applications*
M. Hashem Nehrir and Caisheng Wang

26. *Power Distribution System Reliability: Practical Methods and Applications*
Ali A. Chowdhury and Don O. Koval

27. *Introduction to FACTS Controllers: Theory, Modeling, and Applications*
Kalyan K. Sen, Mey Ling Sen

28. *Economic Market Design and Planning for Electric Power Systems*
James Momoh and Lamine Mili

29. *Operation and Control of Electric Energy Processing Systems*
James Momoh and Lamine Mili

30. *Restructured Electric Power Systems: Analysis of Electricity Markets with Equilibrium Models*
Xiao-Ping Zhang

31. *An Introduction to Wavelet Modulated Inverters*
S.A. Saleh and M.A. Rahman

32. *Control of Electric Machine Drive Systems*
Seung-Ki Sul

33. *Probabilistic Transmission System Planning*
Wenyuan Li

34. *Electricity Power Generation: The Changing Dimensions*
    Digambar M. Tagare
35. *Electric Distribution Systems*
    Abdelhay A. Sallam and Om P. Malik
36. *Practical Lighting Design with LEDs*
    Ron Lenk and Carol Lenk
37. *High Voltage and Electrical Insulation Engineering*
    Ravindra Arora and Wolfgang Mosch
38. *Maintaining Mission Critical Systems in a 24/7 Environment*, Second Edition
    Peter Curtis
39. *Power Conversion and Control of Wind Energy Systems*
    Bin Wu, Yongqiang Lang, Navid Zargari, and Samir Kouro
40. *Integration of Distributed Generation in the Power System*
    Math H. Bollen and Fainan Hassan
41. *Doubly Fed Induction Machine: Modeling and Control for Wind Energy Generation Applications*
    Gonzalo Abad, Jesús López, Miguel Rodrigues, Luis Marroyo, and Grzegorz Iwanski
42. *High Voltage Protection for Telecommunications*
    Steven W. Blume
43. *Smart Grid: Fundamentals of Design and Analysis*
    James Momoh
44. *Electromechanical Motion Devices*, Second Edition
    Paul Krause, Oleg Wasynczuk, and Steven Pekarek
45. *Electrical Energy Conversion and Transport: An Interactive Computer-Based Approach*, Second Edition
    George G. Karady and Keith E. Holbert
46. *ARC Flash Hazard and Analysis and Mitigation*
    J.C. Das
47. *Handbook of Electrical Power System Dynamics: Modeling, Stability, and Control*
    Mircea Eremia and Mohammad Shahidehpour
48. *Analysis of Electric Machinery and Drive Systems*, Third Edition
    Paul C. Krause, Oleg Wasynczuk, S.D. Sudhoff, and Steven D. Pekarek
49. *Extruded Cables for High-Voltage Direct-Current Transmission: Advances in Research and Development*
    Giovanni Mazzanti and Massimo Marzinotto
50. *Power Magnetic Devices: A Multi-Objective Design Approach*
    S.D. Sudhoff
51. *Risk Assessment of Power Systems: Models, Methods, and Applications*, Second Edition
    Wenyuan Li
52. *Practical Power System Operation*
    Ebrahim Vaahedi

53. *The Selection Process of Biomass Materials for the Production of Bio-Fuels and Co-Firing*
    Najib Altawell
54. *Electrical Insulation for Rotating Machines: Design, Evaluation, Aging, Testing, and Repair*, Second Edition
    Greg C. Stone, Ian Culbert, Edward A. Boulter, and Hussein Dhirani
55. *Principles of Electrical Safety*
    Peter E. Sutherland
56. *Advanced Power Electronics Converters: PWM Converters Processing AC Voltages*
    Euzeli Cipriano dos Santos Jr. and Edison Roberto Cabral da Silva
57. *Optimization of Power System Operation*, Second Edition
    Jizhong Zhu
58. *Power System Harmonics and Passive Filter Designs*
    J.C. Das
59. *Digital Control of High-Frequency Switched-Mode Power Converters*
    Luca Corradini, Dragan Maksimoviæ, Paolo Mattavelli, and Regan Zane
60. *Industrial Power Distribution*, Second Edition
    Ralph E. Fehr, III
61. *HVDC Grids: For Offshore and Supergrid of the Future*
    Dirk Van Hertem, Oriol Gomis-Bellmunt, and Jun Liang
62. *Advanced Solutions in Power Systems: HVDC, FACTS, and Artificial Intelligence*
    Mircea Eremia, Chen-Ching Liu, and Abdel-Aty Edris
63. *Operation and Maintenance of Large Turbo-Generators*
    Geoff Klempner and Isidor Kerszenbaum
64. *Electrical Energy Conversion and Transport: An Interactive Computer-Based Approach*
    George G. Karady and Keith E. Holbert
65. *Modeling and High-Performance Control of Electric Machines*
    John Chiasson
66. *Rating of Electric Power Cables in Unfavorable Thermal Environment*
    George J. Anders
67. *Electric Power System Basics for the Nonelectrical Professional*
    Steven W. Blume
68. *Modern Heuristic Optimization Techniques: Theory and Applications to Power Systems*
    Kwang Y. Lee and Mohamed A. El-Sharkawi
69. *Real-Time Stability Assessment in Modern Power System Control Centers*
    Savu C. Savulescu
70. *Optimization of Power System Operation*
    Jizhong Zhu
71. *Insulators for Icing and Polluted Environments*
    Masoud Farzaneh and William A. Chisholm
72. *PID and Predictive Control of Electric Devices and Power Converters Using MATLAB®/Simulink®*
    Liuping Wang, Shan Chai, Dae Yoo, Lu Gan, and Ki Ng

73. *Power Grid Operation in a Market Environment: Economic Efficiency and Risk Mitigation*
Hong Chen

74. *Electric Power System Basics for Nonelectrical Professional*, Second Edition
Steven W. Blume

75. *Energy Production Systems Engineering*
Thomas Howard Blair

76. *Model Predictive Control of Wind Energy Conversion Systems*
Venkata Yaramasu and Bin Wu

77. *Understanding Symmetrical Components for Power System Modeling*
J.C. Das

78. *High-Power Converters and AC Drives*, Second Edition
Bin Wu and Mehdi Narimani

79. *Current Signature Analysis for Condition Monitoring of Cage Induction Motors: Industrial Application and Case Histories*
William T. Thomson and Ian Culbert

80. *Introduction to Electric Power and Drive Systems*
Paul Krause, Oleg Wasynczuk, Timothy O'Connell, and Maher Hasan

81. *Instantaneous Power Theory and Applications to Power Conditioning*, Second Edition
Hirofumi, Edson Hirokazu Watanabe, and Mauricio Aredes

82. *Practical Lighting Design with LEDs*, Second Edition
Ron Lenk and Carol Lenk

83. *Introduction to AC Machine Design*
Thomas A. Lipo

84. *Advances in Electric Power and Energy Systems: Load and Price Forecasting*
Mohamed E. El-Hawary

85. *Electricity Markets: Theories and Applications*
Jeremy Lin and Jernando H. Magnago

86. *Multiphysics Simulation by Design for Electrical Machines, Power Electronics and Drives*
Marius Rosu, Ping Zhou, Dingsheng Lin, Dan M. Ionel, Mircea Popescu, Frede Blaabjerg, Vandana Rallabandi, and David Staton

87. *Modular Multilevel Converters: Analysis, Control, and Applications*
Sixing Du, Apparao Dekka, Bin Wu, and Navid Zargari

88. *Electrical Railway Transportation Systems*
Morris Brenna, Federica Foiadelli, and Dario Zaninelli

89. *Energy Processing and Smart Grid*
James A. Momoh

90. *Handbook of Large Turbo-Generator Operation and Maintenance*, Third Edition
Geoff Klempner and Isidor Kerszenbaum

91. *Advanced Control of Doubly Fed Induction Generator for Wind Power Systems*
Dehong Xu, Dr. Frede Blaabjerg, Wenjie Chen, and Nan Zhu

92. *Electric Distribution Systems*, Second Edition
Abdelhay A. Sallam and Om P. Malik

93. *Power Electronics in Renewable Energy Systems and Smart Grid: Technology and Applications*
Bimal K. Bose

94. *Distributed Fiber Optic Sensing and Dynamic Rating of Power Cables*
Sudhakar Cherukupalli and George J Anders

95. *Power System and Control and Stability,* Third Edition
Vijay Vittal, James D. McCalley, Paul M. Anderson, and A. A. Fouad

96. *Electromechanical Motion Devices: Rotating Magnetic Field-Based Analysis and Online Animations,* Third Edition
Paul Krause, Oleg Wasynczuk, Steven D. Pekarek, and Timothy O'Connell

97. *Applications of Modern Heuristic Optimization Methods in Power and Energy Systems*
Kwang Y. Lee and Zita A. Vale

98. *Handbook of Large Hydro Generators: Operation and Maintenance*
Glenn Mottershead, Stefano Bomben, Isidor Kerszenbaum, and Geoff Klempner

99. *Advances in Electric Power and Energy: Static State Estimation*
Mohamed E. El-hawary

100. *Arc Flash Hazard Analysis and Mitigation,* Second Edition
J.C. Das

101. *Maintaining Mission Critical Systems in a 24/7 Environment,* Third Edition
Peter M. Curtis

102. *Real-Time Electromagnetic Transient Simulation of AC-DC Networks*, First Edition
Venkata Dinavahi and Ning Lin

103. *Probabilistic Power System Expansion Planning with Renewable Energy Resources and Energy Storage Systems,* First Edition
Jaeseok Choi and Kwang Y. Lee

104. *Power Magnetic Devices: A Multi-Objective Design Approach*, Second Edition
Scott D. Sudhoff.

105. *Optimal Coordination of Power Protective Devices with Illustrative Examples,* First Edition
Ali R. Al-Roomi

106. *Resilient Control Architectures and Power Systems*, First Edition
Craig Rieger, Ronald Boring, Brian Johnson, and Timothy McJunkin

107. *Alternative Liquid Dielectrics for High Voltage Transformer Insulation Systems: Performance Analysis and Applications,* First Edition
Edited by U. Mohan Rao, I. Fofana, and R. Sarathi

108. *Power System Protection: Fundamentals and Applications,* First Edition
John Ciufo and Aaron Cooperberg

109. *Soft-Switching Technology for Three-phase Power Electronics Converters*, First Edition
Dehong Xu, Rui Li, Ning He, Jinyi Deng, and Yuying Wu

Printed and bound by CPI Group (UK) Ltd, Croydon, CR0 4YY
17/09/2026
14964252-0001